Anil Kumar, Rituraj Chandrakar, Vikas Dubey and Marta Michalska-Dom

High-Entropy Alloys

Also of Interest

Bulk Metallic Glasses and Their Composites.
Additive Manufacturing and Modeling and Simulation
Muhammad Musaddique Ali Rafique, 2021
ISBN 978-3-11-074721-8, e-ISBN 978-3-11-074723-2

Einführung in die Kristallographie
Joachim Bohm, Detlef Klimm, Manfred Mühlberg and Björn Winkler, 2021
Founded by: Will Kleber
ISBN 978-3-11-046023-0, e-ISBN 978-3-11-046024-7

Multiferroics.
Fundamentals and Applications
Andres Cano, Dennis Meier and Morgan Trassin (Eds.), 2021
ISBN 978-3-11-058097-6, e-ISBN 978-3-11-058213-0

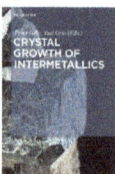

Crystal Growth of Intermetallics
Peter Gille and Yuri Grin (Eds.), 2019
ISBN 978-3-11-049584-3, e-ISBN 978-3-11-049678-9

High-Entropy Alloys

Processing, Alloying Element, Microstructure,
and Properties

Edited by
Anil Kumar, Rituraj Chandrakar, Vikas Dubey,
and Marta Michalska-Domańska

DE GRUYTER

Editors

Dr. Anil Kumar
Department of Mechanical Engineering
Bhilai Institute of Technology
Bhilai House, Durg
Chhattisgarh 491001
India
anilmech2010@gmail.com

Dr. Vikas Dubey
Bhilai Institute of Technology
Abhanpur Road Atal Nagar
Raipur, Chhattisgarh 393661
India
jsvikasdubey.physics@gmail.com

Rituraj Chandrakar
Department of Mechanical Engineering
NMDC DAV Polytechnic, Geedam
Education City, Jawanga
Dakshin Bastar
Dantewada, Chhattisgarh 494441
India
rituraj.chandraker@gmail.com

Dr. Marta Michalska-Domańska
Military Institute of Optoelectronics
Military University of Technology
Kaliskiego 2
00-908 Warsaw
Poland
marta.michalska@wat.edu.pl

ISBN 978-3-11-076944-9
e-ISBN (PDF) 978-3-11-076947-0
e-ISBN (EPUB) 978-3-11-076965-4

Library of Congress Control Number: 2022949432

Bibliographic information published by the Deutsche Nationalbibliothek
The Deutsche Nationalbibliothek lists this publication in the Deutsche Nationalbibliografie;
detailed bibliographic data are available on the internet at http://dnb.dnb.de.

Contents

About the editors

Dr. Anil Kumar is currently serving as an Associate Professor (Mechanical engineering Department) in the BIT, Durg, Chhattisgarh. He has completed his Phd & Post-doc from the National Institute of Technology Raipur. He has published over 30 scientific research articles in the field of Material science & High entropy alloy and also presented various research papers in national and international conferences.

Mr. Rituraj Chandrakar is currently serving as a Lecturer (Mechanical engineering Department) in the NMDC DAV Polytechnic, Dantewada Chhattisgarh. He has over 13 years' experience in teaching and research. He has published over 40 scientific research articles and 3 books in the field of Material science, High entropy alloy, and Supply chain management in various international publications.

Dr. Vikas Dubey is currently serving as an Associate Professor in the BIT, Raipur, Chhattisgarh . He has completed his PhD from the National Institute of Technology, Raipur in 2015. He has received the young scientist award from Chhattisgarh State and has published more than 100 articles in SCI journals and five books.

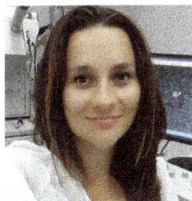

Dr. Michalska-Domanska works at the Military University of Technology in Poland. She is an expert in materials science and focuses on synthesis and characterization of nanomaterials. She is co-author of more than 40 scientific publications and 5 chapters; her H-index is 19.

https://doi.org/10.1515/9783110769470-203

List of contributing authors

Om Prakash
Department of Mechanical Engineering
Jhada Sirha Government Engineering College,
Jagdalpur, Bastar, Chhattisgarh 494001
India
E-mail: omprakash@gecjdp.ac.in

Rituraj Chandrakar
Department of Mechanical Engineering
NMDC DAV Polytechnic, Geedam
Dantewada, Chhattisgarh 494441
And
Department of Mechanical Engineering
National Institute of Technology Karnataka
(NITK)
Surathkal, Mangalore, Karnataka 575025
India

Anil Kumar
Department of Mechanical Engineering
Bhilai Institute of Technology
Durg, Chhattisgarh 491001
India

Marta Michalska-Domańska
Department of Materials Science and
Engineering
Faculty of Advanced Technology and Chemistry
Military University of Technology
Warsaw
Poland

Sheetal Kumar Dewangan
Department of Material Science and Engineering
Ajou University Suwon
South Korea

Sanjay Singh
Department of Mechanical Engineering
CSIT
Durg, Chhattisgarh 491001
India

Manoj Chopkar
Department of Metallurgical and Materials
Engineering
National Institute of Technology
Raipur, Chhattisgarh 492001
India

Rajesh Kumar
Department of Mechanical Engineering
CSIT
Durg, Chhattisgarh 491001
India

Hanuman Reddy Tiyyagura
Rudolfovo - Science and Technology
Centre Novo mesto Podbreznik 15, 8000
Novo mesto Slovenia

Saurabh Chandraker
Department of Mechanical Engineering
National Institute of Technology Karnataka
(NITK)
Surathkal, Mangalore, Karnataka 575025
India

Kundan Lal Sahu
Department of Mechanical Engineering
CSVTU
Bhilai, Chhattisgarh 41001
India
E-mail: saykdn.lal@gmail.com

Saket Kumar
Department of Metallurgical and Materials
Engineering
National Institute of Technology
Raipur, Chhattisgarh 492010
India

https://doi.org/10.1515/9783110769470-204

Ankur Jaiswal
Department of Mechatronics Engineering
Manipal Institute of Technology (MAHE)
Manipal 576104, Karnataka
India

Vikas Dubey
Department of Physics
Bhilai Institute of Technology
Raipur, Chhattisgarh 493661
India

Vikrant Tapas
Department of Mechanical Engineering
NMDC DAV Polytechnic College Geedam
Dantewada, Chhattisgarh
India

Bojanki Naveen
Department of Mechanical Engineering
National Institute of Technology Karnataka
(NITK)
Surathkal, Mangalore, Karnataka 575025
India

Agnivesh Kumar Sinha
Mechanical Engineering Department
Rungta College of Engineering and Technology
Bhilai, Chhattisgarh 490024
India
E-mail: sinhaagnivesh@yahoo.in

Harendra Kumar Narang
Mechanical Engineering Department
National Institute of Technology Raipur
Raipur, Chhattisgarh 492010
India

Somnath Bhattacharya
Mechanical Engineering Department
National Institute of Technology Raipur
Raipur, Chhattisgarh 492010
India

Chandan Pandey
Department of Mechanical Engineering
Indian Institute of Technology Jodhpur
Karwar, Rajasthan 342037
India

Ram Krishna Rathore
Mechanical Engineering Department
Rungta College of Engineering and Technology
Bhilai, Chhattisgarh 490024
India

Vinay Kumar Soni
Mechanical Engineering Department
Columbia Institute Engineering and Technology
Raipur, Chhattisgarh 493111
India

K. Raja Rao
Department of Mechanical Engineering
Lendi Institute of Engineering and Technology
Vizianagaram, Andhra Pradesh 535005
India

Nitin Upadhyay
Department of Mechanical Engineering
Madhav Institute of Technology and Science
Gwalior
Gwalior, Madhya Pradesh 474005
India

Gulab Pamnani
Department of Mechanical Engineering
Malaviya National Institute of Technology
Jaipur, Rajasthan 302017
India

Pankaj Kumar Gupta
Department of Mechanical Engineering
Guru Ghasidas Vishwavidyalaya
Koni, Bilaspur, Chhattisgarh 495009
India

Prem Shankar Sahu
Department of Mechanical Engineering
Bhilai Institute of Technology
Durg, Chhattisgarh 491001
India

K. Raja Rao
Department of Mechanical Engineering
Lendi Institute of Engineering and Technology
Vizianagaram, Andhra Pradesh 535005
India

Satish Pujari
Department of Mechanical Engineering
Lendi Institute of Engineering and Technology
Vizianagaram, Andhra Pradesh 535005
India

Man Mohan
Department of Mechanical Engineering
Rungta College of Engineering and Technology
Bhilai, Chhattisgarh 490024
India
E-mail: manmohan.cimt@gmail.com

Manoj S. Choudhary
Department of Mechanical Engineering
Rungta College of Engineering and Technology
Bhilai, Chhattisgarh 490024
India

Poonam Diwan
Department of Mechanical Engineering
Vishwavidyalaya Engineering College
District – Surguja
Ambikapur, Chhattisgarh 497001
India

Jagesvar Verma
Department of Manufacturing Engineering
National Institute of Advanced
Manufacturing Technology
Near Kanchnatoli, Hatia, Ranchi,
Jharkhand 834003
India

Arun Kumar Sao
Department of Mechanical Engineering
NMDC DAV Polytechnic College
Geedam, Dantewada, Chhattisgarh 494441
India

K. Sridhar
Department of Mechanical Engineering
Lendi Institute of Engineering and Technology
Vizianagaram, Andhra Pradesh 535005
India

Rakshith B Sreesha
Department of Mechanical
Materials and Aerospace Engg. Indian Institute
of Technology Dharwad
Karnataka, India, 580011
rakshith.bs@iitdh.ac.in

Om Prakash, Rituraj Chandrakar, Anil Kumar,
Marta Michalska-Domańska

Chapter 1
Overview of high-entropy alloys

Abstract: New materials and alloys are being developed by using latest technology and manufacturing techniques. Significant progress in alloy system has led to development of special alloys, such as alloys of iron, copper, superalloys, and high-entropy alloys. High-entropy alloys with multiple constituent elements, higher mixing entropy, improved property, and structure make them different from other alloy systems. High- entropy alloy concepts have come into focus after successful development of these alloys, from 2004. Basic concepts, design strategy, phase formation rule, and basic core effects for enhancements of property and structural stability of high-entropy alloys are discussed in this chapter.

Keywords: High-entropy alloys (HEAs), Multicomponent alloys, Multi-principal-element alloys, Configurational entropy, Mixing enthalpy, Phase formation rule, Core effects of high-entropy alloys

1.1 Historical background

The development of new materials has a significant impact on human society. Stone, wood, leather, and bone were among the natural materials used by prehistoric humans throughout the Stone Age. The Stone Age, the Bronze Age, and the Iron Age have been followed by the Steel age, and it has become simpler to extract copper, iron, tin, mercury, and lead from their respective ores. Due to the use of modern technology, the mass production of these materials has developed [1]. There are few uses for pure metals, but there are many for their alloys [2]. Copper alloys were produced by alloying copper mostly with tin, zinc, lead and steels; cast iron is produced by

Om Prakash, Department of Mechanical Engineering, Jhada Sirha Government Engineering College, Jagdalpur, Bastar, Chhattisgarh 494001, India, e-mail: omprakash@gecjdp.ac.in
Rituraj Chandrakar, Department of Mechanical Engineering, NMDC DAV Polytechnic, Geedam, Dantewada, Chhattisgarh 494441, India; Department of Mechanical Engineering, National Institute of Technology Karnataka (NITK), Surathkal, Mangalore 575025, India
Anil Kumar, Department of Mechanical Engineering, Bhilai Institute of Technology, Durg, Chhattisgarh 491001, India
Marta Michalska-Domańska, Department of Materials Science and Engineering, Faculty of Advanced Technology and Chemistry, Military University of Technology, Warsaw, Poland

https://doi.org/10.1515/9783110769470-001

alloying of iron, primarily with carbon. This alloy system is mostly used in our daily life, in transportation, building, weapons, etc. [1, 2].

These days, we have access to a wide range of materials. The **Ashby map**, which is shown in Figure 1.1, provides a comprehensive overview of the evolution in the use of different kinds of materials, over a period of ten millennia. The various kinds of materials, including ceramics, glasses, metals, alloys, polymers, and composites are graphically represented.

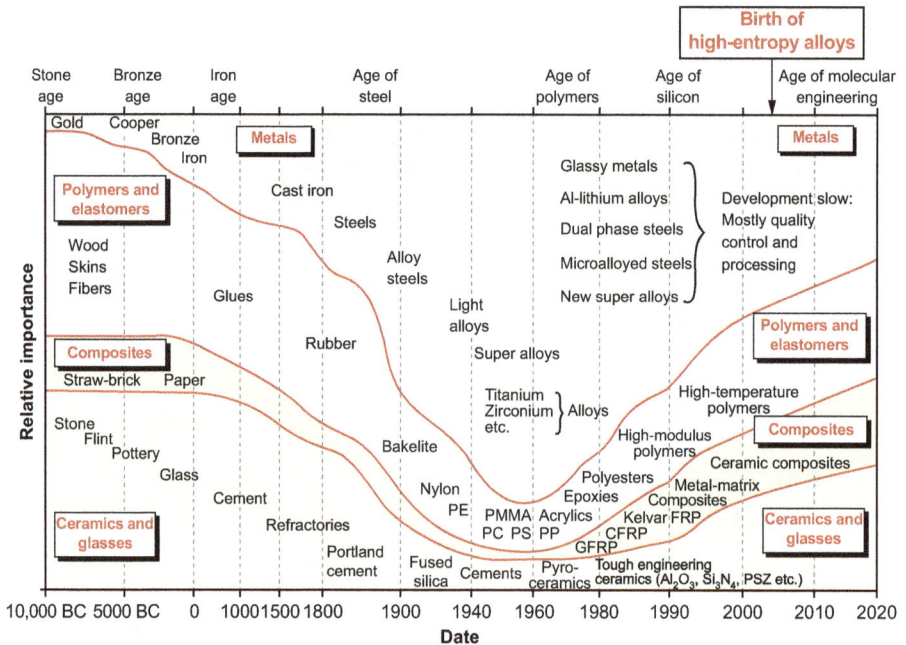

Figure 1.1: The development of HEAs as reported in Advanced Engineering Materials marks a turning point in the historical development of engineering materials [2–4].

Possibly the only researcher whose work on **multicomponent** equimass alloys has been cited in literature is Franz Karl Archard, a German scientist, in the late eighteenth century [5]. His findings demonstrated that the qualities of the alloys are varied but unattractive, using 5–7 elements chosen from the eight commonly occurring elements in the period, including Fe, Cu, Pb, Zn, Sn, Bi, Sb, and As [5]. It is obvious that this is the reason why ancient researchers did not develop and study multielement alloys system [1, 5]. Theoretically, HEAs were considered as early as 1981 [6] and 1996 [7], Professors Jien-Wei Yeh from Taiwan and Brian Cantor from UK have been studying multicomponent alloy system in equal and almost equivalent molar/atomic ratios since 1995 and 1981, respectively. Yeh and his group of researchers created the first **HEAs** in 2004. So, these novel alloys were named "high-entropy alloys (HEAs)" by Prof. Yeh. Exploration of the understudied alloy world was initiated by three major publications: two HEAs/ multicomponent alloy-based publications in 2004 by Jien-Wei Yeh and Brian Cantor, as well as one dependent publication, "Alloyed Pleasures: Multimetallic Cocktails" in 2003, by Prof. S. Ranganathan in India [2, 3, 8, 9]. Through experimental studies of multielement alloy system, a new alloy concept known as "high-entropy alloys (HEAs) or multicomponent alloys" was developed. A timeline of the advancements in HEAs is shown in Figure 1.2. This innovative idea marked a significant turning point in the history of alloy research, and it quickly sparked an international interest from both academics and industry [1].

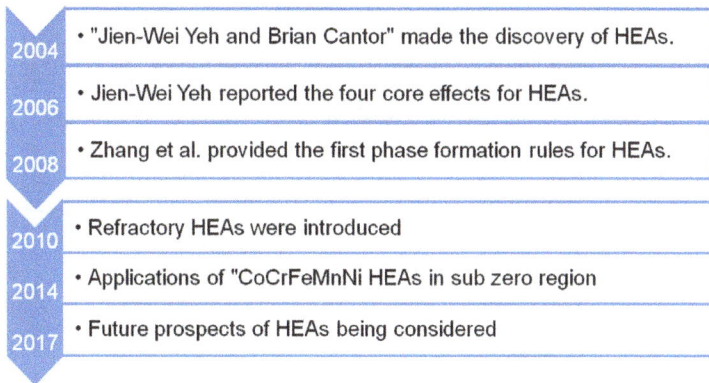

Year	Event
2004	• "Jien-Wei Yeh and Brian Cantor" made the discovery of HEAs.
2006	• Jien-Wei Yeh reported the four core effects for HEAs.
2008	• Zhang et al. provided the first phase formation rules for HEAs.
2010	• Refractory HEAs were introduced
2014	• Applications of "CoCrFeMnNi HEAs in sub zero region
2017	• Future prospects of HEAs being considered

Figure 1.2: Timeline of HEA and related advancements [3, 8, 10–12].

1.2 Research publication progress on high-entropy alloys

Figure 1.3 displays the annual progress of research publication in the HEAs. Up to April 2021, 3,541 research publications based on HEAs had been published cumulatively, according to statistics from the Scopus database. Since 2004, the number of publications has increased consistently each year [13]. The number of scientific research publications on HEAs is shown in Figure 1.3 by year, and it can be seen that this number is gradually increasing [2].

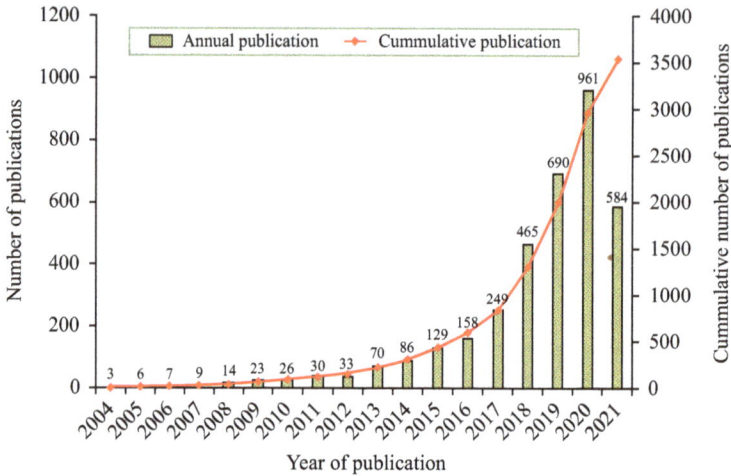

Figure 1.3: Analysis of publication trends for the HEA/MCA between 2004 and April 2021. The information was gathered from the Scopus database [13].

1.3 Basic concept of HEAs

1.3.1 Introduction

There are various methods of manufacturing HEAs, and each HEA has its own unique property and microstructure, and that must be recognized and understood [14]. It is important to focus on the basic concepts and fundamental law related to HEAs, including its origin, basic definition, alloy design strategy, thermodynamics of HEAs, composition notation, phase formation rule, core effects, and application of HEAs [2, 3]. Due to **high configurational entropy**, usually, these alloys are known as HEAs [2]. Due to specific microstructure and unique properties, researches have focused their attention on these alloys [2].

A few advantages, such as great structural stability, high strength, high temperature stability and strength, high hardness, exceptional wear resistance, **creep resistance**, excellent corrosion, and high oxidation resistance at high temperature make HEAs appealing in a range of applications, because some of them are absent from **conventional alloys** [15]. Figure 1.4 shows the yield strength vs. hardness (HV) values for HEAs, which are comparable to BMGs and, often, exceed the standards set by conventional alloys [16].

As seen in Figure 1.5, the strengths of HEAs are frequently higher than those of the majority of metallic alloys. Due to superior **solid solution strengthening**, HEAs have high yield strength, but their ductility is also very impressive. According to Figure 1.4 and Figure 1.5, HEAs appear to provide the best combination of high strength and ductility among all materials [16].

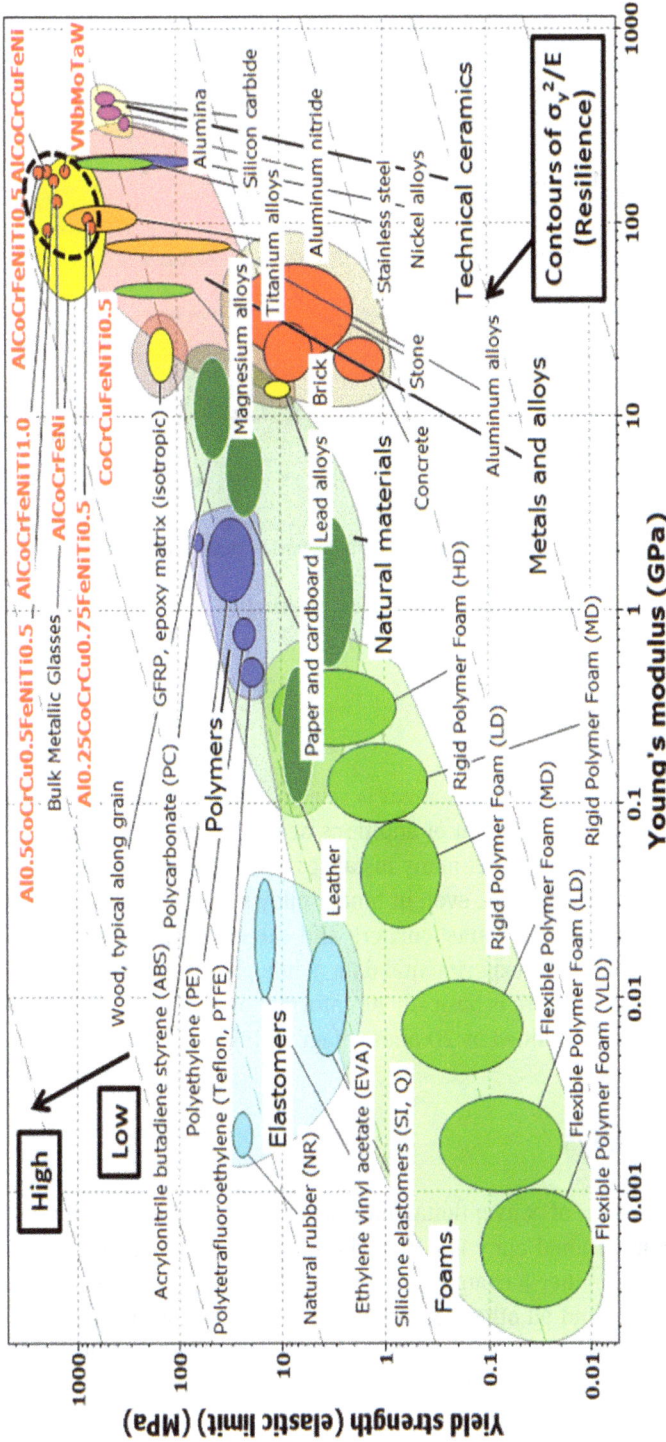

Figure 1.4: An Ashby map of yield strength (elastic limit) vs. Young's modulus for various materials. The upper right black-dashed enclosed ellipse represents the entire HEA family [16].

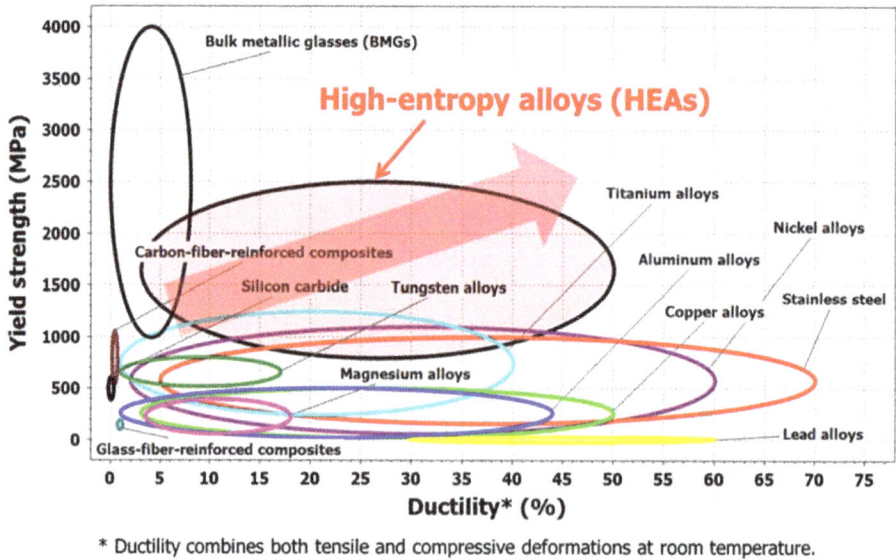

Figure 1.5: An Ashby map of yield strength vs. ductility for various materials [16].

1.4 Alloy design strategy

In conventional alloy design, a principal element is selected, based on its properties, such as Fe, Cu, or Al. Then, minor amount of additions of new elements are made to enhance or add properties. There are not many instances of both elements being employed in approximately equal amounts, even in binary alloy systems, like Pb–Sn solders. Therefore, **ternary phase** diagrams' corners and the edges of binary phase diagrams are well understood from experimental data, while phases towards the centers are much less understood. There is basically no information available for higher-order systems that cannot be explained by 2D phase diagram [8].

1.4.1 Conventional alloys

Since ancient times, the design of conventional and special alloys has focused on one principal element. A conventional alloy is a material made of usually two or more than two elements – that is, either a compound or a solid solution, with one element acting as the major element and all other components acting as minor elements [17]. While alloying materials enhances particular properties for specific applications, pure elements with low impurities are still used in various applications.

Conventional Alloy systems: two or more than two elements, in which there is one principal element, and all other than principal elements are known as minor elements.

$$\text{Conventional Alloy} = A + B + C + ; \ldots$$
$$\text{Principal element} \quad A > 50\%; \ldots$$
$$\text{Minor element} \quad B, C \ldots$$

For example,

$$\text{Steel A} = \text{Fe, B} = \text{Carbon, B} < 2.1\%;$$
$$\text{Cast Iron A} = \text{Fe, B} = \text{Carbon, B} < 6.67\%$$

1.4.2 High-entropy alloys

High-entropy alloys containing at least five principal elements, each with **atomic percentage** between 5 to 35 [2].
- Multi-element systems: 5–13 principal elements
- Equiatomic composition of elements between 5% and 35%
- Any other minor elements <5%

Most of all reported HEAs structures have either the BCC or FCC, as shown in Figure 1.6 [18]. The majority of elements favor a BCC or FCC structure; therefore, this observation is not all that surprising [14].

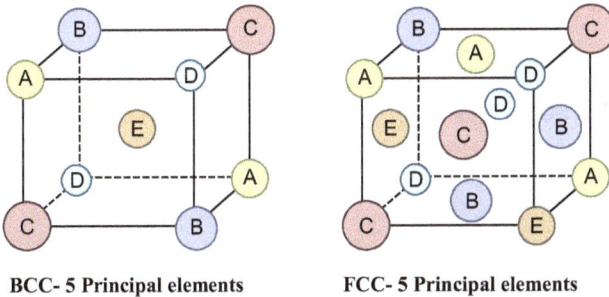

BCC- 5 Principal elements   FCC- 5 Principal elements

Figure 1.6: Diagram of BCC and FCC crystal structure with 5 principal elements [14, 18].
Note: FCC-type HEA solid solution such as CoCrCuFeNi.
BCC-type HEA solid solution such as AlCoCrFeNi.

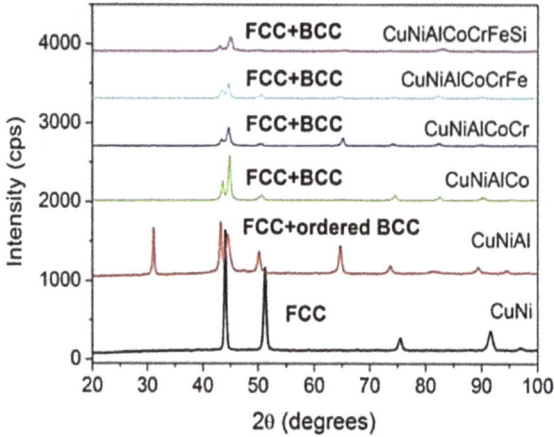

Figure 1.7: The XRD patterns of an alloy created by adding one more element to the preceding one in a series. Every alloy contains FCC or FCC + BCC major phases [15].

A variety of binary to septenary alloys' XRD patterns are displayed in Figure 1.7. There are just two major phases, BCC and FCC, in 3,4,5,6, and 7-component alloys systems. The development of numerous binary and ternary compounds violates expectations when such simple structures are present [15].

1.5 High-entropy alloys are also known by other names [1, 2, 14]

- Multi-principal-component/element alloys
- **Compositionally complex alloys**
- Multicomponent alloy
- Concentrated solid solution alloys
- Equimolar alloys
- Equiatomic ratio alloys

1.6 Composition notation

The chemical formula sequence of the constituent elements can be displayed in a variety of ways, because each HEA's composition does not consist of a single primary element. One typical practice is to arrange the names of the primary elements in alphabetical order, followed by the names of the minor elements, also in a similar alphabetical order, in order to enable simple identification of alloy system. Despite being practical, this has no additional significance. The elements can also be arranged by atomic number or **Mendeleev number** [2]. Atomic ratios or atomic percentages in the subscript position can be used to express the concentration. For example, this rule is followed by the formulas for the two equiatomic alloys, AlCoCrFeNiTi and CoCrCu-FeMnMoNiZr. This criterion is also followed by the formulas for non-equiatomic alloys, as in $Al_{0.5}Co_{1.5}CrFeNi_{1.5}Ti_{0.5}$.

$Co_{17.5}Cr_{8.8}Cu_{8.8}Fe_{26.3}Mn_{8.8}Mo_{5.3}Ni_{17.5}Zr_{7.0}$ in at.% is another way to show these compositions. Additionally, the atomic ratios of the non-equiatomic alloys with minor additions are shown as $Al_{0.5}Co_{1.5}CrFeNi_{1.5}Ti_{0.5}B_{0.1}C_{0.2}$. Due to the fact that molar ratio and atomic ratio are equivalent, molar ratio has also been used in HEA literature. In $Al_{0.5}Co_xCrFeNi_{1.5}Ti_{0.5}B_{0.1}C_{0.2}$, HEAs have five compositions, in which Co is changed from 0.5 to 1.5 (x = 0.5, 0.75, 1.0, 1.25, and 1.5) [2].

1.7 Thermodynamics of HEAs

The fundamental focus of thermodynamics is the interaction between macroscale variables like pressure, temperature, and volume that describe the physical characteristics of substance and heat interaction. Entropy's function in chemical reactions is investigated by chemical thermodynamics. Statistical thermodynamics provide explanations for macroscopic behavior [14, 19].

1.7.1 Entropy

Entropy is a property of the system that can be used to calculate the amount of energy that is available to do work in a process, such as in engines, or energy conversion devices.

Entropy is defined as:

$$dS = \frac{\partial Q}{T} \tag{1.1}$$

where dS is the change in entropy, ∂Q is the heat transfer, and T is the temperature [14].

1.7.2 Gibbs free energy of mixing

Thermodynamically, a system reaches equilibrium when the Gibbs free energy of the system (ΔG_{mix}) *reaches its global minimum* [2].
ΔG_{mix} is defined as:

$$\Delta G_{mix} = \Delta H_{mix} - T\Delta S_{mix} \tag{1.2}$$

where ΔH_{mix} is enthalpy of mixing, ΔS_{mix} is entropy of mixing, T is Temperature, and
ΔH_{mix} and $T\Delta S_{mix}$ are used to determine the free energy for formation of phase of the alloy system. High value of $T\ \Delta S_{mix}$ promotes the solid solution formation of alloy system, because it significantly lowers the ΔG_{mix}, and the HEA is allowed to form more readily [1, 2].

If $\Delta S_{mix} \geq 1.5R$, solid solution occurs in multicomponent system, known as HEAs. This factor stabilizes one single phase rather than multiple phases predicted by Gibbs Phase Rule [14]. If ΔS_{mix} increases, ΔG_{mix} decreases, and this is favorable for solid solution [2].

– These alloys are known as HEAs because they have substantially higher ΔS_{mix} in their random solid solution phases than do conventional alloys. For instance, alloys include CoCrFeNiCu, FeCrMnNiCo, and CuCoNiCrAlFeTiV. The impact of entropy is significantly prominent in HEAs.
– Low Gibbs energy promotes solid solution phases.
– High ΔS_{conf} suppresses ordering, especially at higher temperatures.

1.7.3 Concept of configuration entropy

The statistical mechanics explained the microscopic particles behavior of the system; the statistical mechanics-based definition of entropy was firstly given by **Ludwig Boltzmann** in the 1870s [20, 21]. **Boltzmann's hypothesis** states that "the entropy of a system is linearly related to the logarithm of the frequency of occurrence of a macrostate or, more precisely, the number, w, of possible microstates corresponding to the macroscopic state of a system."

As a result of mixing pure components, alloys have configuration or mixing entropies. From statistical thermodynamics, Boltzmann's equation [1, 22, 23] calculates the ΔS_{conf} of a system as follows:

$$\Delta S_{conf} = k \ln w \tag{1.3}$$

where $k = 1.38 \times 10^{-23}$ J/K is Boltzmann's constant, and the logarithm is taken to be the natural base, e, and w is the number of ways in which the available energy can be mixed or shared among the particles of the system. Thus, the configurational entropy changes per mole for the formation of a solid solution from n elements with X_i mole fraction is expressed as:

$$\Delta S_{conf} = -R \sum_{i=1}^{n} X_i \ln X_i \qquad (1.4)$$

where R is the gas constant, n is the number of components, and X_i is the atomic fraction of component, i. From this it can be seen that alloys in which the components are present in equal proportions will have the highest entropy, and adding additional elements will increase the entropy. A five-component, equiatomic alloy will have **mixing entropy** of 1.61R [2, 3, 10]. The concept of forming equiatomic or nearly equiatomic multiprincipal alloys was independently proposed by Cantor et al. and Yeh et al. [3, 8]. Yeh et al. named these alloys as HEAs by focusing on the fact that ΔS_{conf} of a binary alloy (ΔS_{conf} = $-$ R(X$_A$lnX$_A$ + X$_B$lnX$_B$)) reaches maximum when the alloy elements are in equiatomic ratio (Figure 1.8) and mixing entropies increases when the number of alloying elements (n) in the alloy system increases (Figure 1.9) [3]. Additionally, he emphasized that HEAs have significant mixing entropies and contain both equiatomic and non-equiatomic fractions. He further emphasized that the high ΔH_{mix} in HEAs system would have a significant impact on the "constituent phases, kinetics of phase formation mechanism, lattice strain, and thus properties." In this chapter, "configurational entropy" is mentioned as "mixing entropy/entropy of mixing," yet both these names should be considered as the same. In multicomponent alloys, single solid solution results from the high-entropy effect [3].

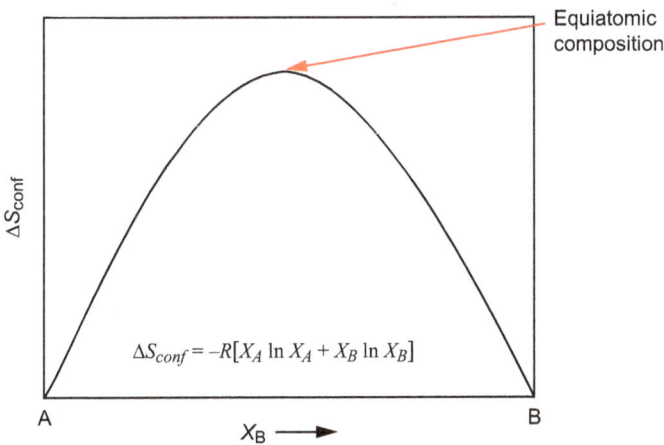

Figure 1.8: In a binary system, configurational entropy ΔS_{conf} reaches its maximum value at equiatomic composition [2, 3].

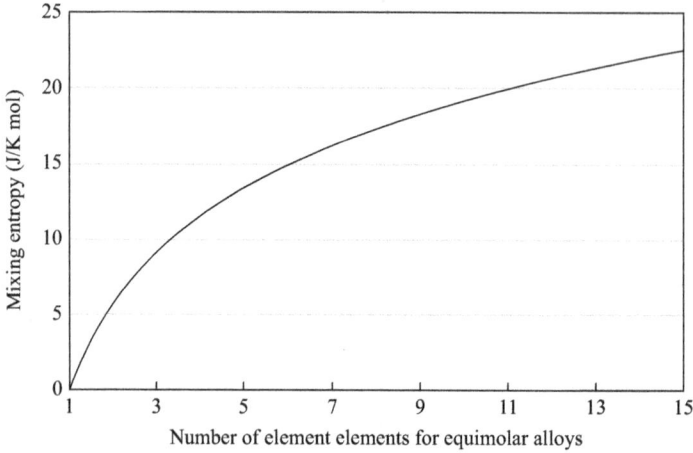

Figure 1.9: Entropy of mixing as a function of the number of elements for equiatomic alloys in the random solution state [2, 3, 13].

For the equiatomic alloys system, the maximum entropy of mixing (ΔS_{mix}) is obtained, and eq. (1.4) can be expressed as:

$$\Delta S_{mix} = R \ln n \tag{1.5}$$

Let us take an equiatomic alloy system in its random solid solution state. Then, its configurational entropy (ΔS_{conf}) per mole is expressed as [3, 24]:

$$\Delta S_{conf} = -k \ln w$$

$$\Delta S_{conf} = -R\left(\frac{1}{n}\ln\frac{1}{n} + \frac{1}{n}\ln\frac{1}{n} + \ldots \frac{1}{n}\ln\frac{1}{n}\right)$$

$$\Delta S_{conf} = -R\ln\frac{1}{n}$$

$$\Delta S_{conf} = R \ln n \tag{1.6}$$

where R is the gas constant, ($R = 8.314$ J/K/mol) and n is number of elements.

Table 1.1: Ideal configurational entropies of equiatomic alloys with constituent elements up to 13 [17].

N	1	2	3	4	5	6	7	8	9	10	11	12	13
ΔS_{conf}	0	0.69R	1.1R	1.39R	1.61R	1.79R	1.95R	2.08R	2.2R	2.3R	2.4R	2.49R	2.57R

Although there are four components of the total mixing entropy such as configurational, vibrational, magnetic dipole, and electronic randomness, configurational entropy dominates the other three [3]. Therefore, to avoid typical calculations to compute the other three components, the configurational entropy frequently serves as a reference for the mixing entropy. The ideal ΔS_{conf} of equiatomic alloys are shown in Table 1.1. As the number of elements (n) increases in the alloy system, the corresponding configurational entropy (ΔS_{conf}) will also increase. As per Richard's rule, the configurational entropy per mole, for change in the phase from a solid state to a liquid state during melting is almost equal to the $1R$ for metals. At high temperatures, the ΔS_{mix} of R per mole is very high in reducing the ΔG_{mix} by the value of RT.

The mixing entropy and enthalpy from chemical bonding are the two key variables that define the equilibrium state, if the impact of strain energy is not considered. The main factor behind the formation of a random solid solution is high mixing entropy and lower mixing enthalpy [2, 17].

1.8 Definition of HEAs

The mixing entropy (ΔS_{mix}) of value 1.5R is quite high to lower the free energy of mixing, and there is a higher possibility to form a random solid solution. From Table 1.1, it is clear that the ΔS_{conf} = 1.61R, for a five-element alloy [17]. Hence, an alloy system with five or at least five principal elements would have higher possibility to form a solid solution. High mixing entropy, in particular, at high working temperatures, promotes the mutual solid solubility of the constituent components and significantly lowers the number of phases. On the basis of above concepts, HEAs are defined by two ways – on the basis of composition and on the basis of configurational entropy [1, 17].

1.8.1 Definition of HEAs on the basis of composition

HEAs are preferentially defined as alloys containing at least five principal elements, each with an atomic percentage between 5 and 35. The atomic percentage of each minor element, if any, is hence less than 5 [2].

The above definition is shown as:

$$n_{major} \geq 5; \quad 5\% \leq X_i \leq 35\%$$

and

$$n_{minor} \geq 0; \quad X_j \leq 5\% \tag{1.7}$$

where n_{major} = number of major elements; $n_{min\,or}$ = number of minor elements (if any); and X_i and X_j are the atomic percentages of major and minor element, respectively [2].

1.8.2 Definition of HEAs on the basis of configurational entropy

HEAs are defined as alloys having configurational entropies at a random state greater than equal to 1.5R, whether they are single phase or multiphase at room temperature [2]. This is defined by the expression:

$$\Delta S_{conf} \geq 1.5\,R \tag{1.8}$$

The high configuration entropy ($\Delta S_{conf} \geq 1.5\,R$) is required to form a single solid solution and prevent the formation of multiphase or **intermetallic compounds**. Therefore, following this guideline is crucial to avoiding HEAs with complex structures and brittleness. It also ensures that the majority of HEAs are feasible to synthesize, investigate, and use. ΔS_{conf} is the only thermodynamic property that rises with the number of primary components. With the help of configurational entropy, we can also define medium- and low-entropy alloy, as 1.5R is a separation limit between high- and medium-entropy alloy, and 1.0R is the separation limit between low alloy and medium alloy entropy, Figure 1.10 shows the type of alloy, based on ΔS_{conf} [1].
For medium-entropy alloys (MEAs):

$$1.0R \leq \Delta S_{conf} \leq 1.5\,R \tag{1.9}$$

For low-entropy alloys (LEAs):

$$\Delta S_{conf} \leq 1.0R \tag{1.10}$$

The definition raises the following query: What is the maximum number of metallic principal elements? The total configurational entropies for 5, 8, 10, 13, 14, 20, and 25 component equiatomic alloys are 1.61R, 2.08R, 2.3R, 2.56R, 2.64R, 3.0R, and 3.22R, respectively. The change in configuration entropy is only 0.07R after increase for each extra element beyond the 13 elements, which is a negligibly small amount. Hence, a feasible number of elements range from 5 to 13 was usually suggested for HEAs [3, 17].

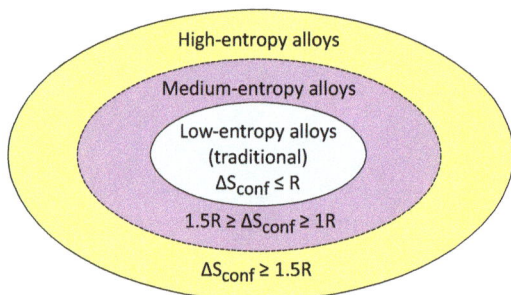

Figure 1.10: Types of alloys based on configurational entropy [17].

As a result, adding more primary elements will not have much of an impact on the high-entropy effect, but they might make managing raw materials or recycling alloys more difficult [2, 17].

1.8.3 Key points on HEAs [1–3, 17, 24]

- HEAs are alloys with five or more than five principal elements, equiatomic, multi-element systems that crystallize as a single phase, despite containing multiple elements with different crystal structures.
- Multielement systems: 5–13 principal elements.
- Equiatomic or near equiatomic composition of elements between 5% and 35%.
- Any other minor elements <5%.
- Stable single phase solid solutions, no precipitates.
- Disordered solid solutions.
- Density varies from 6.7–7.3 gm/cm^3.ΔH_{mix} > – 10 and <5 kJ/mol.
- $\Delta S_{conf} \geq 1.5\,R$.

1.9 Phase formation rule

1.9.1 Introduction

The maximum number of phases in any equilibrium system can be found by using Gibbs' phase rule. Cantor et al. developed a 20-element alloy each with 5 at% Mn, Cr, Fe, Co, Ni, Cu, Ag, W, Mo, Nb, Al, Cd, Sn, Pb, Bi, Zn, Ge, Si, Sb, and Mg in his 2004 work [8]. The phase rule predicted up to 21 phases to exist in equilibrium under constant pressure, although much fewer actually formed. FCC solid solution, predominately containing Fe, Ni, Cr, Co, and Mn was the dominant phase. The outcome of the study was that the only one solid solution phase "FeCrMnNiCo" alloy was successfully developed [8].

A mixture's potential to form a solid solution has traditionally been determined using the **Hume-Rothery rules**. In multicomponent systems, Hume-Rothery rule is a little relaxed, according to research on HEAs. Many elements have different crystal structures, contradicting the rule that solute and solvent atoms of elements must have the same crystalline structure [25].

The performance and applicability of HEAs are significantly influenced by their phase and structure [14]; designing HEAs with specific property requires an understanding of the phase and structure of HEAs. Solid solution, intermetallic, complex, amorphous, and mixtures of these phases are only a few of the various phases found in HEAs. Then, we need to discuss the standard thermodynamic rule and criteria for assessing HEAs and how these standards relate to **crystal structures** [26].

1.9.2 The creation of solid solutions and metallic glass is predicted by thermodynamic factors [2, 10]

The phase formation rule or criteria for HEAs are defined using a thermodynamic factor and parametric method, using physiochemical characteristics such as **enthalpy of mixing** (ΔH_{mix}), entropy of mixing (ΔS_{mix}), atomic size difference (δ), valence electron concentration (VEC), and various other parameters [10] and mismatch entropy (ΔS_σ) were used to define the Bulk Metallic Glass [27].

1.9.2.1 Enthalpy (ΔH_{mix}) versus atomic size difference (δ) parameters

Figure 1.11 shows the phase selection criteria for HEAs and BMGs. The ΔH_{mix} for the multicomponent alloys system can be calculated by [10]:

$$\Delta H_{mix} = \sum_{i=1, i \neq j}^{n} 4 \, \Delta H_{AB}^{mix} X_i X_j \tag{1.11}$$

The atomic size difference (δ) is defined as [1, 10]:

$$\delta = \sqrt{\sum_{i=1}^{n} X_i (1 - d_i / \sum_{j=1}^{n} X_j d_j)^2} \tag{1.12}$$

where ΔH_{AB}^{mix} is a mixing enthalpy of the binary equiatomic alloy, and X_i, d_i, and X_j, d_j are the composition and atomic radii of the ith or jth component, respectively.

Mixing entropy can be shown as follows:

$$\Delta S_{mix} = -R \sum_{i=1}^{n} X_i \ln X_i \tag{1.13}$$

For simple solid solutions phase formation, the parameters satisfying the following criteria, by Zhang et al. [10]:

$$-20 \leq \Delta H_{mix} \leq 5kJ/mol, 12 \leq \Delta S_{mix} \leq 17.5J/Kmol \text{ and } \delta \leq 6.4\%$$

(Figure 1.11) in case of only disordered phases:

$$\Delta H_{mix}(\leq -15kJ/mol) \text{ and a smaller } \delta \leq 6.4\%.$$

Guo et al. analyzed the ΔH_{mix} versus δ plot (Figure 1.11) of various HEAs and found that solid solution formed in the upper left corners of the ΔH_{mix} vs. δ plot [28].

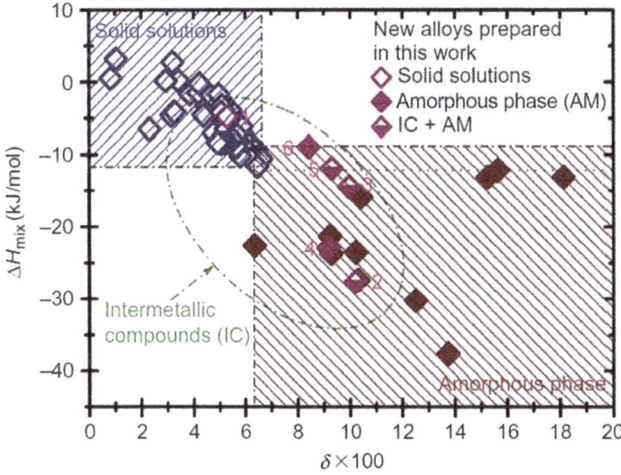

Figure 1.11: A δ - ΔH_{mix} plot explaining the phase selection criteria in HEAs [2, 28].

1.9.2.2 Ω-parameter

The combining effects of ΔS_{mix} and ΔH_{mix} on the phase formation criteria of solid solution were used by one parameter known as Ω-parameter, suggested by Yang and Zhang [1, 29–31].

The Ω-parameter is expressed as:

$$\Omega = \frac{T_m \Delta S_{mix}}{|\Delta H_{mix}|} \tag{1.14}$$

$$T_m = \sum_{i=1}^{n} X_i (T_m)_i \tag{1.15}$$

Color-coded symbols are used to distinguish between solid solutions (S), solid solutions with intermetallic (S + I), intermetallic (I), and amorphous phase (BMGs) (B)) forming multicomponent alloys, where T_m is the mean melting temperature of an alloy system and $(T_m)_i$ is the melting point of the ith element of the alloy. By analyzing the phase formation diagram using the δ and Ωparameter of various previous reported HEA systems (shown in Figure 1.12), the criteria for the formation of solid solution phases, were proposed as: $\Omega \geq 1.1$ and $\delta \leq 6.6\%$. Intermetallic compounds and amorphous phase have higher δ and relatively smaller Ω-parameter, and the value of parameter Ω for BMGs is relatively lesser than intermetallic compounds. Figure 1.13 shows the number of elements, n vs. Ω- parameter plot. It is observed that solid solution occurs at higher Ω-parameter value and higher number of elements (n), while BMGs fall at lower Ω- parameter and smaller number of elements (n) [1].

From the above plot, it is clear that the atomic size difference (δ) is a very significant parameter, for phase selection. The Ω-parameter can be used to replace ΔS_{mix}

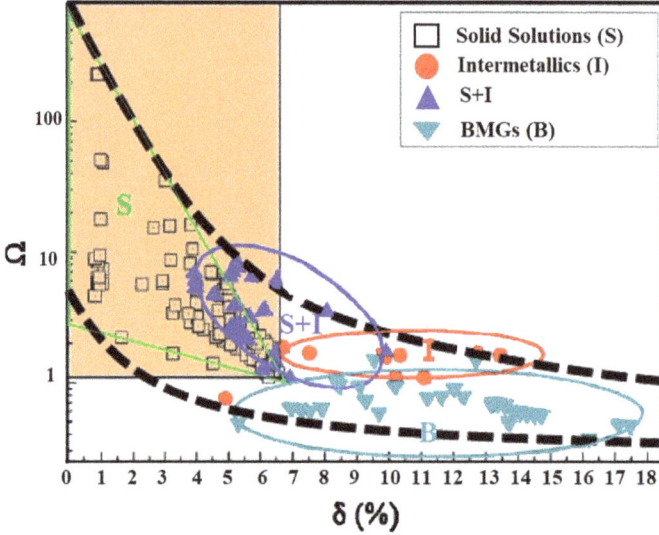

Figure 1.12: Phase selection plot of Ω-parameter versus δ for HEAs as well as BMGs [31, 32].

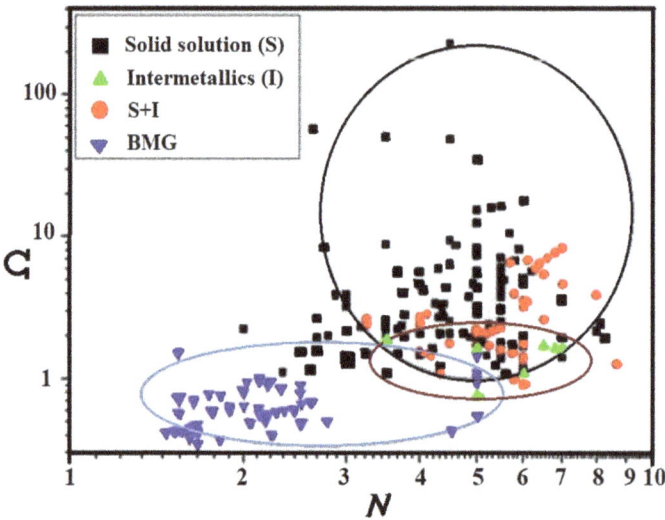

Figure 1.13: Phase selection plot of Ω versus the number of elements (n) for HEAs as well as BMGs [32].

and ΔH_{mix} to estimate the phase formation of HEA. Figure 1.13 shows the phase selection criteria for solid solutions as: $\Omega > 1.1$ and $\delta < 0.066$ [32].

1.9.2.3 Valence electron concentration (VEC)

According to Guo et al., the VEC is a crucial factor in selection of phases in HEAs. From Figure 1.14, it is clear that FCC phase will form when VEC ≥ 8, while BCC will form when the VEC < 6.87 and combined phase (BCC + FCC) coexists, in between. This phase formation criterion is very interesting because FCC formers or stabilizer elements such as Co and Ni have higher VEC values, 9 and 10, respectively, while BCC formers or stabilizer metals, for example, Al and Ti have relatively smaller values, 3 and 4, respectively. This suggests that extra FCC elements are likely to favor FCC phase [33].

VEC is expressed as [33]:

$$VEC = \sum_{i=1}^{n} X_i VEC_i \tag{1.16}$$

where VEC_i is the VEC for the ith element.

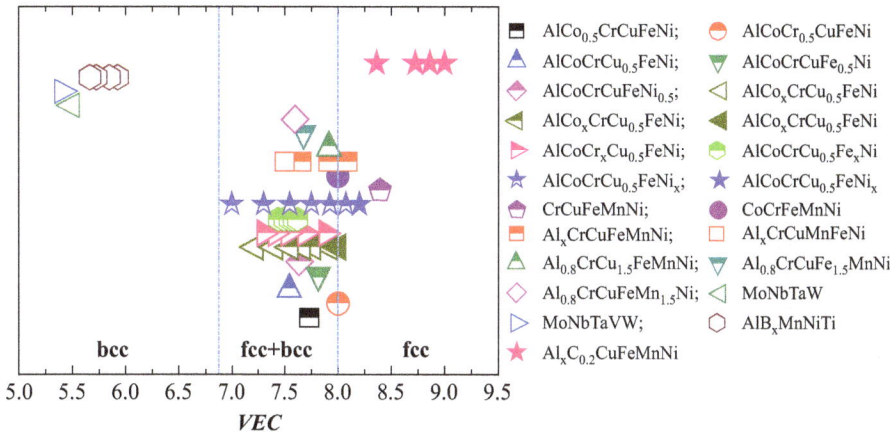

Figure 1.14: Relationship between the phase stability of BCC, FCC and FCC + BCC solid solutions in different HEAs [33].

1.9.2.4 ϕ-parameter

Ye *et al.* [34] has proposed a ϕ-parameter, which states that the total system configurational entropy (ΔS_T) can be given as sum of the configurational entropy, (ΔS_C) and the excessive entropy, (ΔS_E); ($\Delta S_T = \Delta S_C + \Delta S_E$), where ΔS_C and ΔS_E is a function of atomic packing and their sizes. A single dimensionless thermodynamic φ-parameter, is defined by the following equations:

$$\Delta S_H = \frac{|\Delta H_{mix}|}{T_m} \tag{1.17}$$

$$\phi = \frac{\Delta S_C - \Delta S_H}{|\Delta S_E|} \tag{1.18}$$

where ΔS_H is the complementary entropy. The solid solution phases and the thermodynamic criteria are summarized in Figure 1.15. From eq. (1.18), it is expected that the higher ϕ value leads to higher probability of formation of single-phase solid solution in HEAs. HEAs with different solid solution phases are divided by a critical value parameter, $\phi_c = 20$. HEAs are formed in the single-solid solutions phase when $\phi > \phi_c$, and in multiphase as well as **amorphous phase,** when $\phi < \phi_c$. This suggests that ϕ is also an important parameter to find the phase formation of HEAs [35].

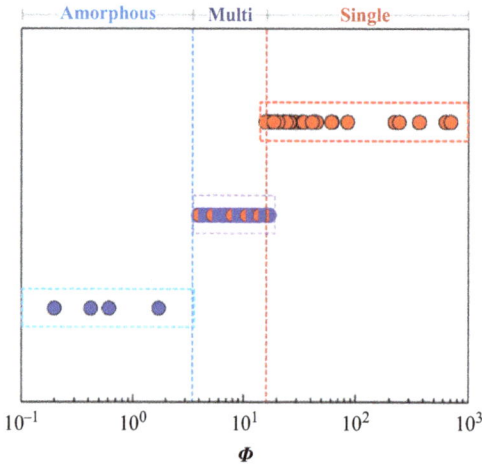

Figure 1.15: The ϕ parameter versus various phases of various HEAs [35].

Figure 1.16 shows the diagram of VEC versus ϕ for various HEAs, with multiphase structure, amorphous phase, and single-phase solid solutions. It is clear that the ϕ parameter and VEC of the HEAs play significant roles in finding the phases. Within the region of the single-phase solid solution ($\phi > \phi c \sim 20$), FCC prevails around VEC of 8.5 ± (1.0), BCC around VEC of 5 ± (0.7), and HCP around VEC of 2.8 ± (0.2) [35].

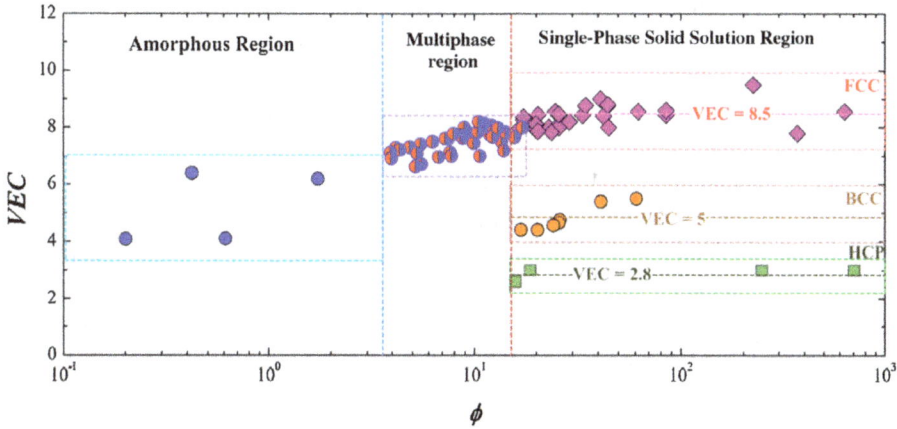

Figure 1.16: Plot of the VEC vs. φ-*parameter* for different HEAs [35].

1.9.2.5 Atomic size difference parameter (γ)

Wang et al. suggested γ-parameter, on the basis of atomic packing mismatch in HEAs. Figure 1.17 shows the plot of γ-δ for various previous studied HEAs. From the plot, it is clear that intermetallic and solid solution phases were formed in the region of 0.04 to 0.08 of δ. It is very complex to distinguish between the phases in this region by using δ -parameter. Yet, γ-parameter can differentiate the solid solution phase from intermetallic; most of all the single-phase solid solution formed when γ < 1.175, and

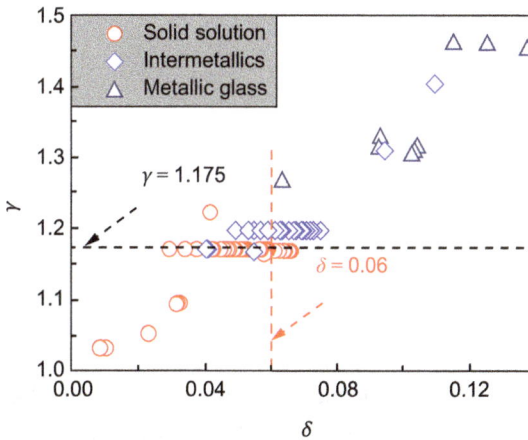

Figure 1.17: Plot of γ vs. δ , where γ = 1.17 is the limiting point for differentiating the solid solution from intermetallic and metallic glass [2, 34].

mostly intermetallic and metallic glass formed when the value $\gamma > 1.175$. Hence, γ-parameter is also a very effective parameter for prediction of phases of HEAs [34].

1.9.2.6 Root mean square (RMS) residual strains

Ye et al. suggested a geometric model that links the alloy's lattice constant to various other parameters, such as atomic size difference and atomic packing density, for the estimation of RMS residual strains around various atomic-sized composition in HEAs. This model can help anticipate the development of amorphous phase in HEAs. With the help of various HEAs data, they plotted a graph between RMS residual strain and elastic energy, as seen in Figure 1.18. They discovered that single-phase solid solution converted into multiphase lattice structure at 5% RMS residual strain and then into amorphous phase at 10% RMS residual strain [36].

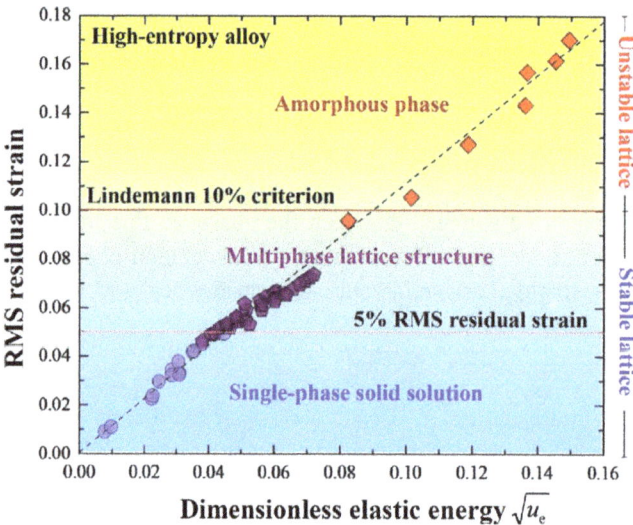

Figure 1.18: The plot between RMS residual strain and the dimensionless elastic energy [2, 36].

1.9.2.7 Parameter Φ

King et al. [37] proposed a parameter, Φ, for predicting the phase formation of HEAs by using the Miedema Model. The plot of parameter Φ versus δ and ΔH_{mix}, respectively is shown in Figure 1.19 (a) and (b); with the help of various HEAs data, it is found that a stable solid solution occurs when $\Phi \geq 1$ [37–39], whereas the condition of $\Phi < 1$ suggests the formation of multiphase [37]. Thermodynamic parameters and proposed phase formation criteria are summarized in Table 1.2.

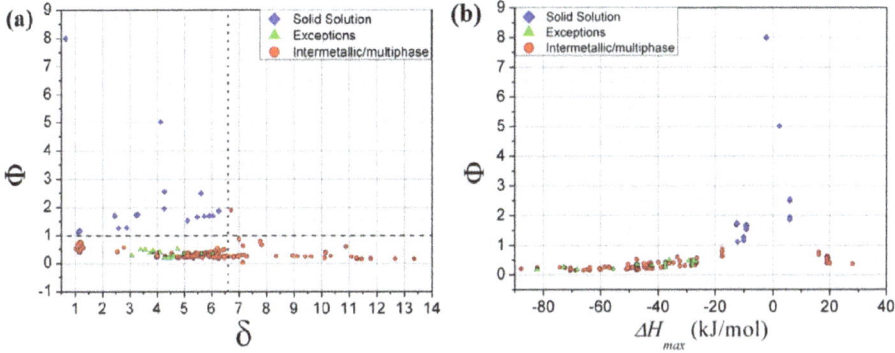

Figure 1.19: The plots of: (a) parameter Φ versus a geometrical parameter, δ and (b) parameter Φ versus ΔH_{max} [37].

Table 1.2: Thermodynamic parameter and proposed criteria of high-entropy alloys.

Thermodynamic parameters	Proposed phase formation criteria			Reference
	Amorphous Phase	Stable solid solution		
		Multiphase	Single phase	
Enthalpy of mixing (ΔH_{mix})	$-49 \le \Delta H_{mix} \le$ 25.5 (kJ mol − 1)	$-22 \ge \Delta H_{mix} \ge 7$ (kJ/mol)		[40]
Entropy of mixing (ΔS_{mix})	$7 \le \Delta S_{mix} \le 16$ (J K − 1 mol − 1)	$11 \ge \Delta \Delta S_{mix} \ge 19.5$ (J/ mol K)		[40]
Valence electron concentration (VEC)	–	FCC: VEC ≥ 8 BCC: VEC ≤ 6.87 BCC + FCC: $6.87 \le$ VEC ≤ 8 HCP: VEC around 3		[40, 41]
Atomic size difference (δ)	$\delta \ge 9\%$	$6.6\% < \delta \le 9\%$	$\delta \le 6.6\%$	[31]
RMS residual strain (εRMS)	εRMS > 10%	5% < εRMS < 10%	εRMS < 5%	[36]
Ω parameter	$\Omega < 1$	$\Omega \ge 1.1$		[31]
Φ Parameter	–	$\Phi < 1$	$\Phi \ge 1$	[37, 42]
ϕ Parameter	When ϕ is too small	$1 < \phi < 20$	$\phi \ge 20$	[43]

1.10 Core effects of high-entropy alloys

It has been found that interactions between the various constituents present in the HEAs system have a significant effect on their structure and characteristics. Four core effects were suggested by Yeh et al. [24]. High-entropy effects in thermodynamics may prevent complex phase formation. Sluggish diffusion effect in kinetics may take longer phase change. Severe lattice distortion could partially change a structure's properties. Due to the interactions between dissimilar atoms and substantial lattice distortion, a cocktail effect is characteristic of the entire property of HEAs [2]. The core effects of HEAs are shown in Figure 1.20 [1].

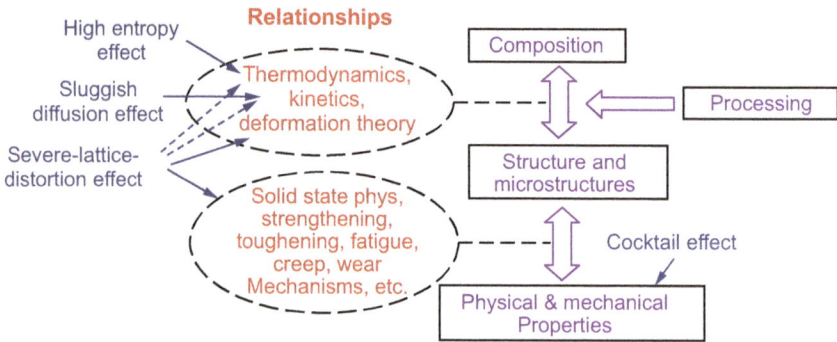

Figure 1.20: The physical metallurgy of HEAs is influenced by four core effects [1].

1.10.1 High-entropy effect

The first significant core effect is the high-entropy effect; this effect minimizes the free energy of mixing and increases the probability of solid solution formation with a simple microstructure [17]. According to maximum entropy production principal (MEPP), high entropy of mixing tends to stabilize the solid solution phases than intermetallic phases [14]. Table 1.3 shows the comparative values of ΔH_{mix}, ΔS_{mix}, and ΔG_{mix}, for elemental phases, compounds, and random solid solution, without taking into account the strain energy effect to mixing enthalpy. It is clear from Table 1.3, that the random solid solution phases have medium negative ΔH_{mix} and high ΔS_{mix}, while compound phases have large negative ΔH_{mix} and very small ΔS_{mix}. At high temperature, solid solution phase formation is more favorable as compared to compound phase. The solid solution phases are favored by a smaller difference in value for unlike pair of atoms. As stated in the literature, a HEA system with components that have small difference in the enthalpy of mixing exhibits a single-phase solid solution, even at high temperatures [44–47]. On the other hand, alloys with a higher difference in mixing enthalpy for unlike pairs of atoms have more than two phases [10]. Researchers have also examined the impact of atomic size difference on various phase formations, and it is reported that smaller size difference, lower mixing enthalpy, and higher entropy of mixing favor the creation of disordered solid solution phases [10, 31, 40].

Table 1.3: Comparative states of ΔH_{mix}, ΔS_{mix}, and ΔG_{mix} for various phases [44].

Comparative states	Elemental phases	Compounds	Intermediate phases	Random solid solutions
ΔH_{mix}	~0	Large -ve	Less large -ve	Medium -ve
ΔS_{mix}	~0	~0	Medium	$\Delta S_{mix} = -R \sum_{i=1}^{n} X_i \ln X_i$
ΔG_{mix}	~0	Large -ve	Larger -ve	Larger -ve

1.10.2 Severe lattice distortion effect

In HEAs, a solid solution phase frequently consists of the entire solute matrix by high-entropy effect, irrespective of the compound's structure [18]. Every atom in the matrix of multiprincipal elements is thus surrounded by various kinds and sizes of atoms and is subject to lattice strain, as seen on the Figure 1.21. A higher lattice distortion is caused by differences in bonding energy, crystal structures among alloy elements, and differences in atomic size because the non-symmetrical atoms that surround an atom site would disturb the atomic position [10, 18, 40, 47, 48]. In conventional alloys system, where most of the solvent atoms have the same kind of atoms as their surrounding, lattice distortion is generally less severe. Figure 1.22 shows an example of a five-component BCC lattice with substantial distortion. A lot of distortion exists in the three-dimensional unit cell [1]. Due to diffuse scattering on warped atomic surfaces, this extreme lattice distortion results in better hardness and a decline in the peak intensity of

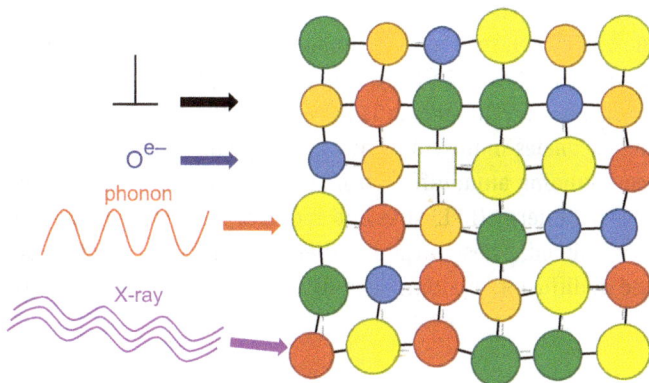

Figure 1.21: Diagrammatic representation of the severely distorted lattice and the different interactions with the X-ray beam, dislocations, electrons, and phonons [1].

1-component alloy   5-components alloy

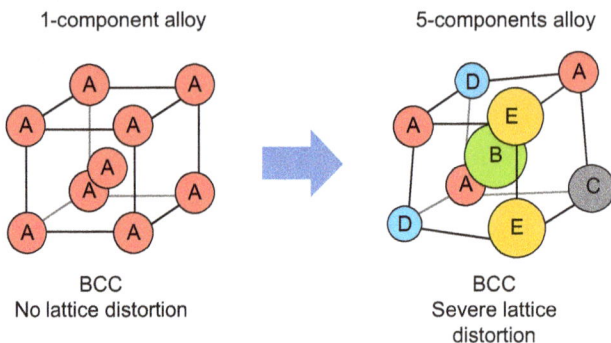

BCC
No lattice distortion

BCC
Severe lattice
distortion

Figure 1.22: Diagram showing that the five-component BCC lattice has significant lattice distortion [2].

X-ray diffraction [18, 24, 48]. In deformed lattices, increased electron and phonon dispersion causes a reduction in electrical and thermal conductivity [24, 49], and increase in hardness due to significant solution hardening in the distorted lattice [11, 24].

1.10.3 Sluggish diffusion effect

Sluggish diffusion effect in HEAs is the effect that has received the most consideration in the literature. It is one of the core effects of HEAs that stands out because it helps improve characteristics by achieving a supersaturated state with very fine precipitates by sluggish grain growth, enhanced recrystallization temperature, decreasing particle coarsening rate, and enhancing creep resistance [44, 50–55]. With these benefits, microstructure and property control might function better. The combination of strength, hardness and toughness, for instance, may be enhanced by very fine grain and precipitation. The lifetime of items used at high temperatures might be extended through improved creep resistance [1, 2]. By diffusion-controlled phase transformation, it is predicted that slow diffusion may have an impact on the nucleation, growth, and distribution of newly formed phases. In HEA, it is very complex to differentiate among solvent and solute atoms, and the matrix is taken to be the entire solute matrix. The atoms around the vacancies in the solute atoms occupy the vacancies during sluggish diffusion. Each surrounding atom differs from the others, which causes fluctuations in the Lattice Potential Energy (LPE) sites and requires greater activation energy for diffusion to be completed. These large numbers of LPE sites found in the HEAs causes the hindrance in diffusion of atoms, leading to sluggish diffusion [44, 47].

1.10.4 Cocktail effect

The term "multimetallic cocktails" was first proposed by Prof. S. Ranganathan in 2003, in his article "Alloyed Pleasures: Multimetallic Cocktails" [9]. According to the cocktail effect, the characteristics of the alloy are influenced by both the individual elements' characteristics as well as their interactions [56]. Since HEAs have five or more principal elements, the solid solution that results can have one, two, or more phases. The properties that are produced are a combination of the effects of all the constituent parts as well as lattice distortion. Each phase of the HEA solid solution can therefore be thought of as an atomic composite. By choosing components that have the properties needed in alloys for certain purposes, this core effect can be utilized [57]. The cocktail effect is used in HEAs to highlight the non-linear impact of various factors on property enhancement. The characteristics of HEA are thought to be a combination of the fundamental properties of the elements as well as the results of their interactions with one another. As a result, the cocktail effect can be thought of as a multiphase cocktail effect, ranging from atomic to microscale. These characteristics are the outcome of a non-linear synergistic interaction, in which certain characteristics derive from the synergistic combination of various elements [44]. High strength, good plasticity, high magnetization, high electric resistance, and low coercivity are all outcomes of this beneficial cocktail effect. Thus, it is essential for a HEAs designer to comprehend the essential elements involved, before selecting appropriate compositions and processes [2]. Due to the "cocktail effect," HEAs are stronger than the average composing elements [1, 2].

References

[1]   Y., G. M. C., Y. J. W., L. P. K., and Z. Y. Zhang, *High-Entropy Alloys*. Cham: Springer International Publishing, 2016. doi: 10.1007/978-3-319-27013-5.

[2]   B. S., Y. J. W., R. S., and B. P. P. Murty, "High-entropy alloys," Elsevier (2nd edition), ISBN:978-0-12-816067-1, 2019.

[3]   J.-W. Yeh, et al., "Nanostructured high-entropy alloys with multiple principal elements: Novel alloy design concepts and outcomes," *Advanced Engineering Materials*, vol. 6, no. 5, pp. 299–303, May 2004, doi: 10.1002/adem.200300567.

[4]   M. F. Ashby and D. Cebon, "Materials selection in mechanical design," *Le Journal de Physique IV*, vol. 03, no. C7, pp. C7-1-C7-9, Nov. 1993, doi: 10.1051/jp4:1993701.

[5]   C. S. Smith, *Four Outstanding Researches in Metallurgical History*. Baltimore: American Society for Testing and Materials, 1963.

[6]   A. J. B. Vincent and B. Cantor, "Part II thesis," University of Sussex, 1981.

[7]   K. H. Huang and J. W. Yeh, "A study on multicomponent alloy systems containing equal-mole elements [M.S. thesis]," Hsinchu: National Tsing Hua University 1, 1996.

[8]   B. Cantor, I. T. H. Chang, P. Knight, and A. J. B. Vincent, "Microstructural development in equiatomic multicomponent alloys," *Materials Science and Engineering: A*, vol. 375–377, pp. 213–218, Jul. 2004, doi: 10.1016/j.msea.2003.10.257.

[9]   S. Ranganathan, "Alloyed pleasures: Multimetallic cocktails," *Current Science*, vol. 85, no. 10, pp. 1404–1406, Nov. 2003.

[10] Y. Zhang, Y. J. Zhou, J. P. Lin, G. L. Chen, and P. K. Liaw, "Solid-solution phase formation rules for multi-component alloys," *Advanced Engineering Materials*, vol. 10, no. 6, pp. 534–538, Jun. 2008, doi: 10.1002/adem.200700240.

[11] O. N. Senkov, G. B. Wilks, D. B. Miracle, C. P. Chuang, and P. K. Liaw, "Refractory high-entropy alloys," *Intermetallics (Barking)*, vol. 18, no. 9, pp. 1758–1765, Sep. 2010, doi: 10.1016/j.intermet.2010.05.014.

[12] B. Gludovatz, A. Hohenwarter, D. Catoor, E. H. Chang, E. P. George, and R. O. Ritchie, "A fracture-resistant high-entropy alloy for cryogenic applications," *Science*, vol. 345, no. 6201, pp. 1153–1158, Sep. 2014, doi: 10.1126/science.1254581.

[13] A. D. Akinwekomi and F. Akhtar, "Bibliometric mapping of literature on high-entropy/multicomponent alloys and systematic review of emerging applications," *Entropy*, vol. 24, no. 3, pp. 329, Feb. 2022, doi: 10.3390/e24030329.

[14] Y. Zhang, et al., "Microstructures and properties of high-entropy alloys," *Progress in Materials Science*, vol. 61, pp. 1–93, Apr. 2014, doi: 10.1016/j.pmatsci.2013.10.001.

[15] Z. Tang, et al., "Aluminum alloying effects on lattice types, microstructures, and mechanical behavior of high-entropy alloys systems," *JOM*, vol. 65, no. 12, pp. 1848–1858, Dec. 2013, doi: 10.1007/s11837-013-0776-z.

[16] M.-H. Tsai and J.-W. Yeh, "High-entropy alloys: A critical review," *Materials Research Letters*, vol. 2, no. 3, pp. 107–123, Jul. 2014, doi: 10.1080/21663831.2014.912690.

[17] J.-W. Yeh, "Alloy design strategies and future trends in high-entropy alloys," *JOM*, vol. 65, no. 12, pp. 1759–1771, Dec. 2013, doi: 10.1007/s11837-013-0761-6.

[18] J.-W. Yeh, et al., "Formation of simple crystal structures in Cu-Co-Ni-Cr-Al-Fe-Ti-V alloys with multiprincipal metallic elements," *Metallurgical and Materials Transactions A*, vol. 35, no. 8, pp. 2533–2536, 2014.

[19] T. L. Hill, *An Introduction to Statistical Thermodynamics*. Courier Corporation, 1986.

[20] C. A. Gearhart, "Einstein before 1905: The early papers on statistical mechanics," *American Journal of Physics*, vol. 58, no. 5, pp. 468–480, May 1990, doi: 10.1119/1.16478.

[21] A. R. Ruffa, "Thermal potential, mechanical instability, and melting entropy," *Physical Review B*, vol. 25, no. 9, pp. 5895–5900, May 1982, doi: 10.1103/PhysRevB.25.5895.

[22] D. R. Gaskell and D. E. Laughlin, *Introduction to the Thermodynamics of Materials*, 3rd ed. Taylor & Francis Ltd, 2017.

[23] S. RA, *Thermodynamics of Solids*, 2nd ed. John Wiley and Sons, 1972.

[24] J. W. Yeh, "Recent progress in high-entropy alloys," *Annales de Chimie: Science Des Materiaux*, vol. 31, no. 6, pp. 633–648, Nov. 2006, doi: 10.3166/acsm.31.633-648.

[25] F. Otto, Y. Yang, H. Bei, and E. P. George, "Relative effects of enthalpy and entropy on the phase stability of equiatomic high-entropy alloys," *Acta Materialia*, vol. 61, no. 7, pp. 2628–2638, Apr. 2013, doi: 10.1016/j.actamat.2013.01.042.

[26] X. Wang, W. Guo, and Y. Fu, "High-entropy alloys: Emerging materials for advanced functional applications," *Journal of Chemistry A Materials*, vol. 9, no. 2, pp. 663–701, 2021, doi: 10.1039/D0TA09601F.

[27] A. Takeuchi, K. Amiya, T. Wada, K. Yubuta, W. Zhang, and A. Makino, "Entropies in alloy design for high-entropy and bulk glassy alloys," *Entropy*, vol. 15, no. 12, pp. 3810–3821, Sep. 2013, doi: 10.3390/e15093810.

[28] S. Guo, Q. Hu, C. Ng, and C. T. Liu, "More than entropy in high-entropy alloys: Forming solid solutions or amorphous phase," *Intermetallics (Barking)*, vol. 41, pp. 96–103, Oct. 2013, doi: 10.1016/j. intermet.2013.05.002.

[29] X. Yang, S. Y. Chen, J. D. Cotton, and Y. Zhang, "Phase stability of low-density, multiprincipal component alloys containing aluminum, magnesium, and lithium," *JOM*, vol. 66, no. 10, pp. 2009–2020, Oct. 2014, doi: 10.1007/s11837-014-1059-z.

[30] Y. Zhang, X. Yang, and P. K. Liaw, "Alloy design and properties optimization of high-entropy alloys," *JOM*, vol. 64, no. 7, pp. 830–838, Jul. 2012, doi: 10.1007/s11837-012-0366-5.

[31] X. Yang and Y. Zhang, "Prediction of high-entropy stabilized solid-solution in multi-component alloys," *Materials Chemistry and Physics*, vol. 132, no. 2–3, pp. 233–238, Feb. 2012, doi: 10.1016/j. matchemphys.2011.11.021.

[32] Y. Zhang, et al., "Guidelines in predicting phase formation of high-entropy alloys," *MRS Communications*, vol. 4, no. 2, pp. 57–62, Jun. 2014, doi: 10.1557/mrc.2014.11.

[33] S. Guo, C. Ng, J. Lu, and C. T. Liu, "Effect of valence electron concentration on stability of FCC or BCC phase in high-entropy alloys," *Journal of Applied Physics*, vol. 109, no. 10, pp. 103505, May 2011, doi: 10.1063/1.3587228.

[34] Z. Wang, Y. Huang, Y. Yang, J. Wang, and C. T. Liu, "Atomic-size effect and solid solubility of multicomponent alloys," *Scripta Materialia*, vol. 94, pp. 28–31, Jan. 2015, doi: 10.1016/j. scriptamat.2014.09.010.

[35] Y. F. Ye, Q. Wang, J. Lu, C. T. Liu, and Y. Yang, "High-entropy alloy: Challenges and prospects," *Materials Today*, vol. 19, no. 6, pp. 349–362, Jul. 2016, doi: 10.1016/j.mattod.2015.11.026.

[36] Y. F. Ye, C. T. Liu, and Y. Yang, "A geometric model for intrinsic residual strain and phase stability in high entropy alloys," *Acta Materialia*, vol. 94, pp. 152–161, Aug. 2015, doi: 10.1016/j. actamat.2015.04.051.

[37] D. J. M. King, S. C. Middleburgh, A. G. McGregor, and M. B. Cortie, "Predicting the formation and stability of single phase high-entropy alloys," *Acta Materialia*, vol. 104, pp. 172–179, Feb. 2016, doi: 10.1016/j.actamat.2015.11.040.

[38] S. Fang, X. Xiao, L. Xia, W. Li, and Y. Dong, "Relationship between the widths of supercooled liquid regions and bond parameters of Mg-based bulk metallic glasses," *Journal of Non-crystalline Solids*, vol. 321, no. 1–2, pp. 120–125, Jun. 2003, doi: 10.1016/S0022-3093(03)00155-8.

[39] V. K. Soni, S. Sanyal, K. R. Rao, and S. K. Sinha, "A review on phase prediction in high entropy alloys," *Proceedings of the Institution of Mechanical Engineers, Part C: Journal of Mechanical Engineering Science*, vol. 235, no. 22, pp. 6268–6286, Nov. 2021, doi: 10.1177/09544062211008935.

[40] S. GUO and C. T. LIU, "Phase stability in high entropy alloys: Formation of solid-solution phase or amorphous phase," *Progress in Natural Science: Materials International*, vol. 21, no. 6, pp. 433–446, Dec. 2011, doi: 10.1016/S1002-0071(12)60080-X.

[41] Y. F. Ye, Q. Wang, J. Lu, C. T. Liu, and Y. Yang, "High-entropy alloy: Challenges and prospects," *Materials Today*, vol. 19, no. 6, pp. 349–362, Jul. 2016, doi: 10.1016/j.mattod.2015.11.026.

[42] R. Feng, et al., "Design of light-weight high-entropy alloys," *Entropy*, vol. 18, no. 9, pp. 333, Sep. 2016, doi: 10.3390/e18090333.

[43] Y. F. Ye, Q. Wang, J. Lu, C. T. Liu, and Y. Yang, "Design of high entropy alloys: A single-parameter thermodynamic rule," *Scripta Materialia*, vol. 104, pp. 53–55, Jul. 2015, doi: 10.1016/j. scriptamat.2015.03.023.

[44] S. Yadav, K. Biswas, and A. Kumar, "Spark plasma sintering of high entropy alloys," In: *Spark Plasma Sintering of Materials*, Springer International Publishing, 2019, pp. 539–571. doi: 10.1007/978-3-030-05327-7_19.

[45] F. Otto, A. Dlouhý, C. Somsen, H. Bei, G. Eggeler, and E. P. George, "The influences of temperature and microstructure on the tensile properties of a CoCrFeMnNi high-entropy alloy," *Acta Materialia*, vol. 61, no. 15, pp. 5743–5755, Sep. 2013, doi: 10.1016/j.actamat.2013.06.018.

[46] O. N. Senkov, J. M. Scott, S. V. Senkova, D. B. Miracle, and C. F. Woodward, "Microstructure and room temperature properties of a high-entropy TaNbHfZrTi alloy," *Journal of Alloys and Compounds*, vol. 509, no. 20, pp. 6043–6048, May 2011, doi: 10.1016/j.jallcom.2011.02.171.

[47] K. Y. Tsai, M. H. Tsai, and J. W. Yeh, "Sluggish diffusion in Co–Cr–Fe–Mn–Ni high-entropy alloys," *Acta Materialia*, vol. 61, no. 13, pp. 4887–4897, Aug. 2013, doi: 10.1016/J.ACTAMAT.2013.04.058.

[48] J. W. Yeh, S. Y. Chang, Y. der Hong, S. K. Chen, and S. J. Lin, "Anomalous decrease in X-ray diffraction intensities of Cu–Ni–Al–Co–Cr–Fe–Si alloy systems with multi-principal elements," *Materials Chemistry and Physics*, vol. 103, no. 1, pp. 41–46, May 2007, doi: 10.1016/J.MATCHEMPHYS.2007.01.003.

[49] Y. F. Kao, S. K. Chen, T. J. Chen, P. C. Chu, J. W. Yeh, and S. J. Lin, "Electrical, magnetic, and Hall properties of AlxCoCrFeNi high-entropy alloys," *Journal of Alloys and Compounds*, vol. 509, no. 5, pp. 1607–1614, Feb. 2011, doi: 10.1016/J.JALLCOM.2010.10.210.

[50] C. Y. Hsu, C. C. Juan, W. R. Wang, T. S. Sheu, J. W. Yeh, and S. K. Chen, "On the superior hot hardness and softening resistance of AlCoCrxFeMo0.5Ni high-entropy alloys," *Materials Science and Engineering: A*, vol. 528, no. 10–11, pp. 3581–3588, Apr. 2011, doi: 10.1016/J.MSEA.2011.01.072.

[51] C.-Y. Hsu, J.-W. Yeh, S.-K. Chen, and -T.-T. Shun, "Wear resistance and high-temperature compression strength of FCC CuCoNiCrAl0.5Fe alloy with boron addition," *Metallurgical and Materials Transactions A*, vol. 35, no. 5, pp. 1465–1469, May 2004, doi: 10.1007/s11661-004-0254-x.

[52] W. H. Liu, Y. Wu, J. Y. He, T. G. Nieh, and Z. P. Lu, "Grain growth and the Hall-Petch relationship in a high-entropy FeCrNiCoMn alloy," *Scripta Materialia*, vol. 68, no. 7, pp. 526–529, Apr. 2013, doi: 10.1016/j.scriptamat.2012.12.002.

[53] O. N. Senkov, G. B. Wilks, J. M. Scott, and D. B. Miracle, "Mechanical properties of Nb25Mo25Ta25W25 and V20Nb20Mo20Ta20W20 refractory high entropy alloys," *Intermetallics (Barking)*, vol. 19, no. 5, pp. 698–706, May 2011, doi: 10.1016/j.intermet.2011.01.004.

[54] C. W. Tsai, Y. L. Chen, M. H. Tsai, J. W. Yeh, T. T. Shun, and S. K. Chen, "Deformation and annealing behaviors of high-entropy alloy Al0.5CoCrCuFeNi," *Journal of Alloys and Compounds*, vol. 486, no. 1–2, pp. 427–435, Nov. 2009, doi: 10.1016/J.JALLCOM.2009.06.182.

[55] M. H. Tsai, J. W. Yeh, and J. Y. Gan, "Diffusion barrier properties of AlMoNbSiTaTiVZr high-entropy alloy layer between copper and silicon," *Thin Solid Films*, vol. 516, no. 16, pp. 5527–5530, Jun. 2008, doi: 10.1016/J.TSF.2007.07.109.

[56] C. Suryanarayana, "Mechanical alloying: A critical review," *Materials Research Letters*, vol. 10, no. 10, pp. 619–647, Oct. 2022, doi: 10.1080/21663831.2022.2075243.

[57] W. Ji, et al., "Alloying behavior and novel properties of CoCrFeNiMn high-entropy alloy fabricated by mechanical alloying and spark plasma sintering," *Intermetallics (Barking)*, vol. 56, pp. 24–27, Jan. 2015, doi: 10.1016/j.intermet.2014.08.008.

Anil Kumar, Sheetal Kumar Dewangan, Sanjay Singh, Manoj Chopkar, Rakshith B Sreesha
Chapter 2
Classification of processing routes

Abstract: There are various synthesis and processing routes for producing high-entropy alloys (HEAs). Based on the constituent elements that are mixed, the processing routes and synthesis techniques are classified into liquid mixing, solid-state mixing, and gas-state mixing. Processing routes mostly affect the property and structural stability of HEAs. Each route and synthesis technique has its own significant advantages, disadvantages, and limitations. Based on the requirements, proper and suitable selection of processing routes and synthesis techniques is important. In this chapter, the classification of processing routes, synthesis techniques, and significant advantages and disadvantages along with certain limitations are discussed in detail.

Keywords: High-entropy alloys (HEAs), Synthesis and processing route, Mechanical alloying, Spark Plasma Sintering (SPS)

2.1 Introduction

Several processing methods have been used for the synthesis of high-entropy alloys (HEAs). Melting and Castings (liquid state), powder metallurgy (solid state), and films (gaseous state) are the three major categories into which the processing routes can be divided. Melting and casting with uniform and non-uniform cooling rates are used to produce HEAs [1], which includes arc melting, vacuum induction melting, laser-enhanced net shape and laser cladding (LC). **Vacuum arc melting** (VAM) and **vacuum induction melting** (VIM) are the two most often used melt processing methods [1].

The powder metallurgy route mostly includes **mechanical alloying** (MA) and consequent consolidation process [2]. The primary method for producing bulk sintered items in the solid state has been MA, followed by sintering [1]. Powders with high

Anil Kumar, Department of Mechanical Engineering, Bhilai Institute of Technology, Durg, Chhattisgarh 491001, India, e-mail: anilmech2010@gmail.com
Sheetal Kumar Dewangan, Department of Material Science and Engineering, Ajou Univercity Suwon, 16499, South Korea
Sanjay Singh, Department of Mechanical Engineering, CSIT, Durg, Chhattisgarh 491001, India
Manoj Chopkar, Department of Metallurgical and Materials Engineering, National Institute of Technology, Raipur, Chhattisgarh 492001, India
Rakshith B Sreesha, Department of Mechanical, Materials and Aerospace Engg. Indian Institute of Technology Dharwad, Karnataka, India, 580011, e-mail: rakshith.bs@iitdh.ac.in

https://doi.org/10.1515/9783110769470-002

melting points can be effectively combined using **Spark Plasma Sintering** (SPS), which can create compacts with densities as high as 99.6% [3]. **Ball milling** is a popular technique for creating powders with nanocrystalline structures.

We can also synthesize the elements from the vapor or gaseous state deposition of HEAs. This processing route includes Sputtering, Pulse Laser Deposition (PLD), Molecular Beam Epitaxy (MBE), and Atomic Layer Deposition (ALD) to synthesize the films on substrates [2]. HEA nanoparticles with remarkable compositional homogeneity were created by Carbothermal Shock (CTS) synthesis [4].

2.2 Classification of processing routes

To create HEAs, a variety of conventional industrial and innovative laboratory techniques have been applied. Three main processing routes have been described in Figure 2.1, which includes several HEAs production processes [2].

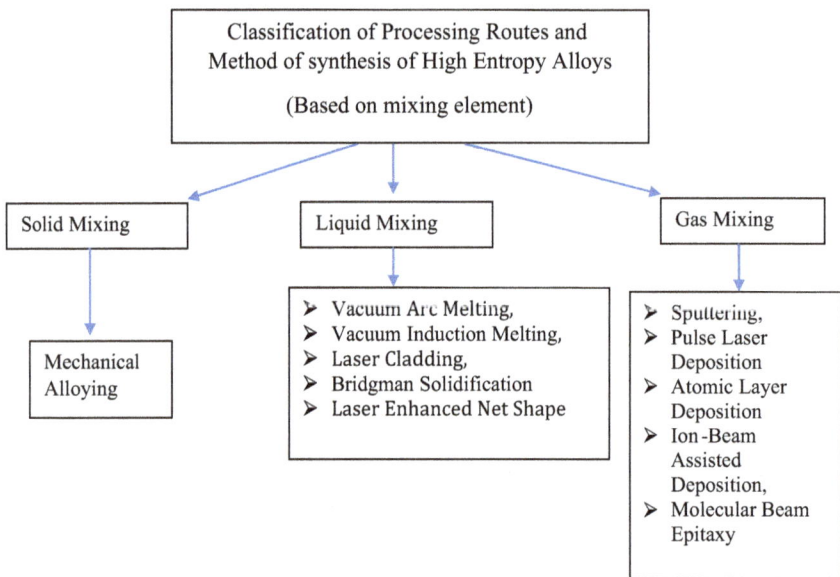

Figure 2.1: Processing route and synthesis methods of HEAs [2].

2.3 Advantages and disadvantages/limitations of each route

Processing routes	Synthesis technique	Advantage	Disadvantage/limitation
Solid-state processing route	Mechanical alloying (MA)	– MA is capable of synthesizing a wide range of metastable phases such as supersaturated solid solution, metallic glass (amorphous alloys), etc. [5]. – Alloys having a positive heat of mixing have a tendency to phase segregate in the liquid and solid condition, which complicates melting and makes it difficult to achieve homogeneity in the melt. However, MA has none of these issues, because it is a fully solid-state synthesis technique and does not include melting [5]. – MA increases the solid solution solubility limits in alloy systems and even in immiscible systems [5].	– When exposed to high temperatures, the microstructure developed by MA can change, and it is typically metastable [1]. – It is challenging to measure the peak position accurately, because the XRD peaks in mechanically alloyed powders are large and diffuse [5]. – The peak positions may shift as a result of lattice distortion and the existence of stacking faults in the MA milled powder [5]. – There is considerable risk that the condition in which the powders are milled could contaminate the powder alloys [1, 2, 5].

(continued)

(continued)

Processing routes	Synthesis technique	Advantage	Disadvantage/limitation
Liquid metallurgy route	Vacuum arc melting	– The most favored method for melting HEAs is arc melting, because it can reach high temperatures that are adequate to melt the majority of the component metals [1]. – Time, money, and energy savings with less effort [6].	– It is challenging to maintain compositional stability in the HEA due to the evaporation that occurs in low boiling point metals [1]. – Dendritic and interdendritic developments are two limitations of segregation at high temperatures [7]. – One of the limitations faced in this technique is the heterogeneous microstructure produced by the slower rate of solidification [1]. – The mechanical properties of HEAs may also be negatively impacted by a number of unavoidable as-cast flaws, such as elemental segregation, inhibition of equilibrium phases, macroscopic residual stresses, cracks, and porosities [2].
	Vacuum induction melting [8]	– Flexibility due to small batch sizes. – Rapid program changes for various steel and alloy types. – Simple operation. – Low oxidation losses of alloying elements. – Proper temperature control. – Low emission of dust that pollutes the environment. – High vapor pressures for the removal of undesirable trace elements. – Removal of dissolved gases like nitrogen and hydrogen.	– Elements having high vapor pressures, like manganese, can cause issues when alloyed. Additionally, vacuum induction melting is more expensive than electric arc and other melting.

Bridgman solidification technique (BST) [9]	– BST can be utilized efficiently to control the microstructure and enhance the properties of HEAs as compared to conventional casting. – Since the solid is free to expand because it is not in contact with the crucible on top during expansion, there is less stress. – Greater mixing in the melt is caused by thermal convection (flows caused by temperature differential).	– Steady furnace temperature makes it challenging to maintain a constant crystal growth [10]. – Small scale of production.
Laser cladding [11]	– High rates of heating and cooling. – Low temperature effect on the substrate. – Small and homogeneous grains are present. – A strong bonding strength is attained when the coating is coupled with the metallurgical matrix. – It can be up to several millimeters thick.	– Various faults are produced easily. – It is clear that the cladding layer is susceptible to cracking.
LENS	– High-resolution features can be produced using this procedure, and dimensional accuracy of a product can be controlled [12–15]. – Compared to traditional manufacturing methods like casting, LENS machines produce smaller melt-pools, enabling faster rates of solidification and cooling [15–17].	– LENS considerably increases residual stress, which can decrease part accuracy or even cause part collapse during deposition [15, 18]. – In LENS, the inconsistent heating and cooling rate might result from non-uniformity in the macro- and microstructures with regard to the height of the construct [15]. – Its post-processing requirements, the uneven component surface finish, and the components' distortion are the result of residual stresses [19].

(continued)

(continued)

Processing routes	Synthesis technique	Advantage	Disadvantage/limitation
Gaseous-state processing route	Magnetron sputtering [11]	– There is a slow increase in substrate temperature and a rapid film deposition. – The film is well structured and consistent throughout. – Good adherence to the substrate. – The parameters enable flexible control of the effectiveness and thickness of the film.	– There is little usage of the target. – There are restrictions on film thickness.
	Pulse Laser Deposition (PLD) [20]	– Adaptable technique (any material). – Consistent evaporation. – Rapid deposition (10 s nm/min). – Clean process. – A high-energy plume. – Reactive gases such as oxygen. – Wide variety of gas pressures.	– Relatively high cost. – Mainly associated with lasers; picosecond and femtosecond pulsed lasers are still not widely used in laboratories or in industry. – Extremely complex models hinder theory-based improvements.
	Ion Beam–Assisted Deposition (IBAD) [21]	– The IBAD technique has the advantage that the ion flux can be typically measured via a "Faraday cup ion-collector," and the atom flow can be quantified via a mass deposition monitor, like a Quartz crystal monitoring system (QCM).	– Reactive deposition cannot be done using plasma activation methods, and the equipment prices are substantially greater than for other ion plating process. – Difficult to scale up (low yield).

Method	Advantages	Disadvantages
Atomic Layer Deposition [22]	– High-quality films because the thickness of the film is controlled, very good repeatability, enhanced film density, ultra-thin films. – Excellent thickness homogeneity over a large area, and smooth surface coating. – Low temperature and stress, smooth deposition for sensitive substrates, and superior adhesion.	– Duration of the chemical reactions. – High rate of material waste. – Extremely high rate of energy loss. – The intensiveness of the ALD process. – Emission of nanoparticles.
Molecular Beam Epitaxy [23]	– Auto-doping and auto-diffusion are both reduced when thin layers are prepared using the MBE technique. – No chemical interactions take place during the process.	– It is challenging to maintain a very low pressure. – In comparison to CVD, this method is highly expensive.
Carbothermal shock (CTS) synthesis [4]	– This method of synthesis offers: (i) universality, (ii) tunability, (iii) possible scalability, and (iv) opens up a huge field for synthesis of nanoparticles of different HEAs. – The CTS method can be used to mix practically any metallic combination uniformly, since its maximum temperature (2,000 to 3,000 K) is much higher than the temperature at which any metal salt decomposes (i.e., generality).	– Multimetal homogenous dispersions are challenging to produce, especially at large catalyst loadings (>10 wt%).

(continued)

(continued)

Processing routes	Synthesis technique	Advantage	Disadvantage/limitation
	High-Pressure Torsion	– The fact that exceptionally high shear strain may be achieved with HPT in a relatively simple manner is its most significant benefit. In many materials, >1,000 is present without issue [24]. – HPT allows for defined continuous strain variation. The HPT method takes place at close to room temperature and does not require further consolidation procedures, in contrast to other mechanical alloying methods [25, 26]. – It is generally known that using HPT processing to add a significant amount of dislocations and vacancies effectively improves atomic mobility and elemental mixing [27, 28].	– HPT sample size restrictions and upscaling [24].

References

[1]   B. S. Murty, J. W. Yeh, and S. Ranganathan, "Synthesis and processing," In: *High Entropy Alloys*, Elsevier, 2014, pp. 77–89. doi: 10.1016/b978-0-12-800251-3.00005-5.

[2]   Y., G. M. C., Y. J. W., L. P. K., and Z. Y. Zhang, *High Entropy Alloys*. Cham: Springer International Publishing, 2016. doi: 10.1007/978-3-319-27013-5.

[3]   S. Fang, W. Chen, and Z. Fu, "Microstructure and mechanical properties of twinned Al0.5CrFeNiCo0.3C0.2 high entropy alloy processed by mechanical alloying and spark plasma sintering," *Materials & Design (1980–2015)*, vol. 54, pp. 973–979, Feb. 2014, doi: 10.1016/j.matdes.2013.08.099.

[4]   Y. Yao, et al., "Carbothermal shock synthesis of high-entropy-alloy nanoparticles," *Science (1979)*, vol. 359, no. 6383, pp. 1489–1494, Mar. 2018, doi: 10.1126/science.aan5412.

[5]   C. Suryanarayana, "Mechanical alloying: A critical review," *Materials Research Letters*, vol. 10, no. 10, pp. 619–647, Oct. 2022, doi: 10.1080/21663831.2022.2075243.

[6]   S. K. Dewangan, A. Mangish, S. Kumar, A. Sharma, B. Ahn, and V. Kumar, "A review on high-temperature applicability: A milestone for high entropy alloys," *Engineering Science and Technology, An International Journal*, p. 101211, Jul. 2022, doi: 10.1016/j.jestch.2022.101211.

[7]   O. Maulik, D. Kumar, S. Kumar, S. K. Dewangan, and V. Kumar, "Structure and properties of lightweight high entropy alloys: A brief review," *Materials Research Express*, vol. 5, no. 5, p. 052001, May 2018, doi: 10.1088/2053-1591/aabbca.

[8]   Handbook Committee, Ed., *Melting, Vacuum Induction. "ASM Handbook 2008'"*, vol. 15. Casting. ASM International, 2008.

[9]   P. S. Dutta, "Bulk growth of crystals of III–V compound semiconductors," In: *Comprehensive Semiconductor Science and Technology*, Elsevier, 2011, pp. 36–80. doi: 10.1016/B978-0-44-453153-7.00090-0.

[10] P. Han, J. Tian, and W. Yan, "Bridgman growth and properties of PMN–PT-based single crystals," In: *Handbook of Advanced Dielectric, Piezoelectric and Ferroelectric Materials*, Elsevier, 2008, pp. 3–37. doi: 10.1533/9781845694005.1.3.

[11] Y. Zhang, "Preparation methods of high-entropy materials," In: *High-Entropy Materials*, Singapore: Springer Singapore, 2019, pp. 65–75. doi: 10.1007/978-981-13-8526-1_3.

[12] W. E. Frazier, "Metal additive manufacturing: A review," *Journal of Materials Engineering and Performance*, vol. 23, no. 6, pp. 1917–1928, Jun. 2014, doi: 10.1007/s11665-014-0958-z.

[13] B. Vandenbroucke and J. Kruth, "Selective laser melting of biocompatible metals for rapid manufacturing of medical parts," *Rapid Prototyping Journal*, vol. 13, no. 4, pp. 196–203, Aug. 2007, doi: 10.1108/13552540710776142.

[14] I. Yadroitsev, L. Thivillon, P. Bertrand, and I. Smurov, "Strategy of manufacturing components with designed internal structure by selective laser melting of metallic powder," *Applied Surface Science*, vol. 254, no. 4, pp. 980–983, Dec. 2007, doi: 10.1016/j.apsusc.2007.08.046.

[15] M. Izadi, A. Farzaneh, M. Mohammed, I. Gibson, and B. Rolfe, "A review of laser engineered net shaping (LENS) build and process parameters of metallic parts," *Rapid Prototyping Journal*, vol. 26, no. 6, pp. 1059–1078, Apr. 2020, doi: 10.1108/RPJ-04-2018-0088.

[16] V. K. Balla, S. Banerjee, S. Bose, and A. Bandyopadhyay, "Direct laser processing of a tantalum coating on titanium for bone replacement structures," *Acta Biomaterialia*, vol. 6, no. 6, pp. 2329–2334, Jun. 2010, doi: 10.1016/j.actbio.2009.11.021.

[17] V. K. Balla, S. Bodhak, S. Bose, and A. Bandyopadhyay, "Porous tantalum structures for bone implants: Fabrication, mechanical and in vitro biological properties," *Acta Biomaterialia*, vol. 6, no. 8, pp. 3349–3359, Aug. 2010, doi: 10.1016/j.actbio.2010.01.046.

[18] K. V. Wong and A. Hernandez, "A review of additive manufacturing," *ISRN Mechanical Engineering*, vol. 2012, pp. 1–10, Aug. 2012, doi: 10.5402/2012/208760.

[19] S. A. Kumar and R. V. S. Prasad, "Basic principles of additive manufacturing: Different additive manufacturing technologies," In: *Additive Manufacturing*, Elsevier, 2021, pp. 17–35. doi: 10.1016/B978-0-12-822056-6.00012-6.

[20] R. Eason, *Pulsed Laser Deposition of Thin Films: Applications-led Growth of Functional Materials*. John Wiley & Sons, 2007.

[21] D. M. Mattox, "Ion plating and ion beam-assisted deposition," In: *Handbook of Physical Vapor Deposition (PVD) Processing*, Elsevier, 2010, pp. 301–331. doi: 10.1016/B978-0-8155-2037-5.00009-5.

[22] P. O. Oviroh, R. Akbarzadeh, D. Pan, R. A. M. Coetzee, and T.-C. Jen, "New development of atomic layer deposition: Processes, methods and applications," *Science and Technology of Advanced Materials*, vol. 20, no. 1, pp. 465–496, Dec. 2019, doi: 10.1080/14686996.2019.1599694.

[23] R. F. Farrow, *Molecular Beam Epitaxy: Applications to Key Materials*. Elsevier, 1995.

[24] R. Pippan, S. Scheriau, A. Hohenwarter, and M. Hafok, "Advantages and limitations of HPT: A review," *Materials Science Forum*, vol. 584–586, pp. 16–21, Jun. 2008, doi: 10.4028/www.scientific.net/MSF.584-586.16.

[25] A. P. Zhilyaev and T. G. Langdon, "Using high-pressure torsion for metal processing: Fundamentals and applications," *Progress in Materials Science*, vol. 53, no. 6, pp. 893–979, Aug. 2008, doi: 10.1016/j.pmatsci.2008.03.002.

[26] K. Edalati and Z. Horita, "A review on high-pressure torsion (HPT) from 1935 to 1988," *Materials Science and Engineering: A*, vol. 652, pp. 325–352, Jan. 2016, doi: 10.1016/j.msea.2015.11.074.

[27] J. Čížek, et al., "The development of vacancies during severe plastic deformation," *Materials Transactions*, vol. 60, no. 8, pp. 1533–1542, Aug. 2019, doi: 10.2320/matertrans.MF201937.

[28] J. González-Masís, J. M. Cubero-Sesin, A. Campos-Quirós, and K. Edalati, "Synthesis of biocompatible high-entropy alloy TiNbZrTaHf by high-pressure torsion," *Materials Science and Engineering: A*, vol. 825, p. 141869, Sep. 2021, doi: 10.1016/j.msea.2021.141869.

Kundan Lal Sahu, Saket Kumar, Ankur Jaiswal, Vikas Dubey

Chapter 3
Powder metallurgy route

Abstract: This chapter offers a discussion on the powder metallurgy route for the production of high-entropy alloys (HEAs). In the powder metallurgy route, the constitutional elements are mixed in their solid state. The selection of the elemental powder composition and the mixing of this elemental powder are also important and are mostly done by mechanical alloying. Mechanical alloying is a solid-state powder mixing and processing technique. Solid powder particles are synthesized in a variety of alloy phases in a high-energy ball mill. In order to attain homogeneous microstructures in HEAs, the powder metallurgy process, which includes mechanical alloying and the consequent consolidation process – spark plasma sintering, hot processing, cold processing, and sintering process, is efficient in terms of time, material, and energy.

Keywords: High-entropy alloys (HEAs), Powder metallurgy route, Mechanical alloying (MA), Powder atomization technique, Spark Plasma Sintering (SPS), Cold pressing (CP) process, Hot pressing process, Sintering process

3.1 Introduction

Ingot metallurgy is used to create the majority of HEAs. Powder metallurgy (PM) has recently emerged as an attractive method for further advancing the features of high-entropy alloys, with the potential to expand the area of **"nanostructures"** in HEAs, and also enhance some of the features of these alloys. In this chapter, powder metallurgy (PM) methods for HEAs are addressed along with some prospective techniques to increase the utilization of powders as raw materials. Most of the research on HEAs has primarily focused on developing processing technologies based on ingot metallurgy. When more than five different types of elements are present, including some that have high melting points, they must be completely melted and solidified with adequate solubility, while preventing segregation. Melting has proven to be a highly effective procedure. However, ingot metallurgy offers significant challenges in cases

Kundan Lal Sahu, Department of Mechanical Engineering, CSVTU, Bhilai, Chhattisgarh 41001, India, e-mail: saykdn.lal@gmail.com

Saket Kumar, Department of Metallurgical and Materials Engineering, National Institute of Technology, Raipur, Chhattisgarh 492010, India

Ankur Jaiswal, Department of Mechatronics Engineering, Manipal Institute of Technology (MAHE), Manipal, Karnataka 576104, India

Vikas Dubey, Department of Physics, Bhilai Institute of Technology, Raipur, Chhattisgarh 493661, India

https://doi.org/10.1515/9783110769470-003

where a complex alloy composition is required, which may limit its potential for developing HEAs. In this context, PM has shown great promise as a method for producing HEAs. PM is a forming process that allows for high compositional accuracy. It may entirely avoid segregation and provide excellent control on the microstructure (including nanocrystalline) [1]. Although the first publications on HEAs were published in 2004, it took a few years for papers using PM to appear [2–4]. PM has two further advantages over other processing techniques: (i) It is applicable for using metals with varying densities, such as when creating lightweight HEAs [5], and (ii) it can be used when several metals with higher melting points are associated with the synthesis of the HEA, as in the case of the so-called **refractory HEAs** [6]. Due to the growing interest in PMHEAs, there is a gradual increase in the number of articles published over the years [7].

In the last decade, material scientists have used a variety of ways to synthesize HEAs, including melting and **casting** (liquid state), **powder metallurgy** (solid state), and gaseous state processing route. In order to attain homogeneous microstructures in HEAs, the PM process, which includes mechanical alloying and the consequent consolidation process by SPS, is commonly used. The MA and the spark plasma sintering processes are efficient in terms of time, material, and energy. However, contamination from the grinding media makes PM synthesis of HEAs difficult [8]. Mechanical alloying with a "**high-energy ball mill**" is the most common method of solid-state processing. This procedure creates powders that can later be treated using SPS. This process enables the production of alloys such as AlLiMgScTi, which would be difficult or impossible to create by casting [9, 10].

3.2 Powder metallurgy route: chemical compositions of HEAs

Ingot metallurgy was used to create the first HEAs, which had at least five major elements in equimolar quantities. Various compositions are available in the literature, most of which are based on transition metals in the 3d. To be designated a HEA, such complex compositions must meet a set of basic criteria set forth by Yeh et al. [11]. The prerequisites are four core effects – high-entropy effects, severe lattice distortion effect, sluggish diffusion effect and **cocktail effects** [12]. Three main categories of elements emerge from an analysis of how frequently certain alloying elements have been used (Figure 3.1), which is illustrated in Figure 3.2. The most frequently used elements are transition metals (TM): "Fe, Ni, Cr, and Co." Another group of alloying elements, also connected to transition metals (apart from Al), is usually less used: Ti, Cu, Mn, Al [1]. If we use the definitions of HEA set forth by Yeh et al., we can see that we would require at least five elements [13]. According to the distribution illustrated in Figure 3.2, the majority of the alloys tested are made up of four metals from the first group and, mostly, one or two from the second. V, Mo, Ta, Zn, Nb, W, and Zr belong to a third attractive group of elements that are employed less frequently. The third group of elements, with the exclusion of Zn, comprises refractory metals that play a key part in the refractory HEAs [6]. Following the initial development of HEAs, based mostly on 3d, TM refractory HEAs were produced, with very particular and distinct characteristics that span a broad spectrum of applications. Finally, certain examples of the utilization of elements, such as Si and Mg, are recounted, resulting in a fourth group, shown in Figure 3.2.

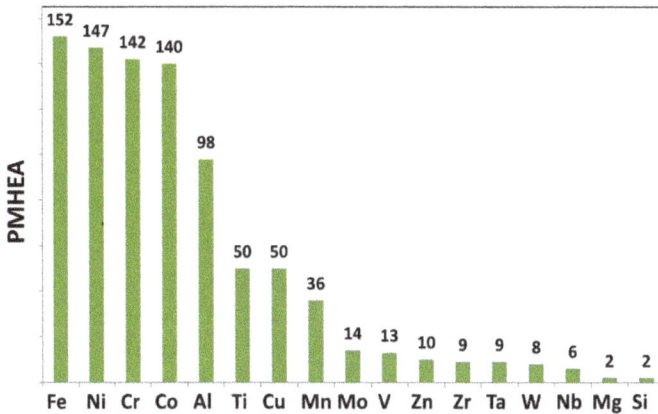

Figure 3.1: In the investigated PM HEAs, the distribution (frequency) of the alloying elements [1].

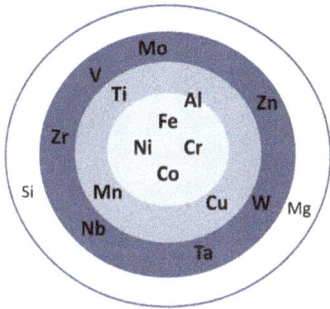

Figure 3.2: In the HEAs studied, the main alloying elements [1].

3.3 Solid-state techniques or conventional powder metallurgy route

The **solid-state processing** (SSP) approach, which involves combining various alloying elements at room temperature and then consolidating them at a higher temperature, is used to make the majority of polycrystalline materials. The SSP technique can be utilized to manufacture HEAs, as shown in Figure 3.3. The SSP route includes powder manufacturing, powder mixing, powder consolidation into bulk samples, SPS, cold compaction, hot pressing and many other secondary activities [14].

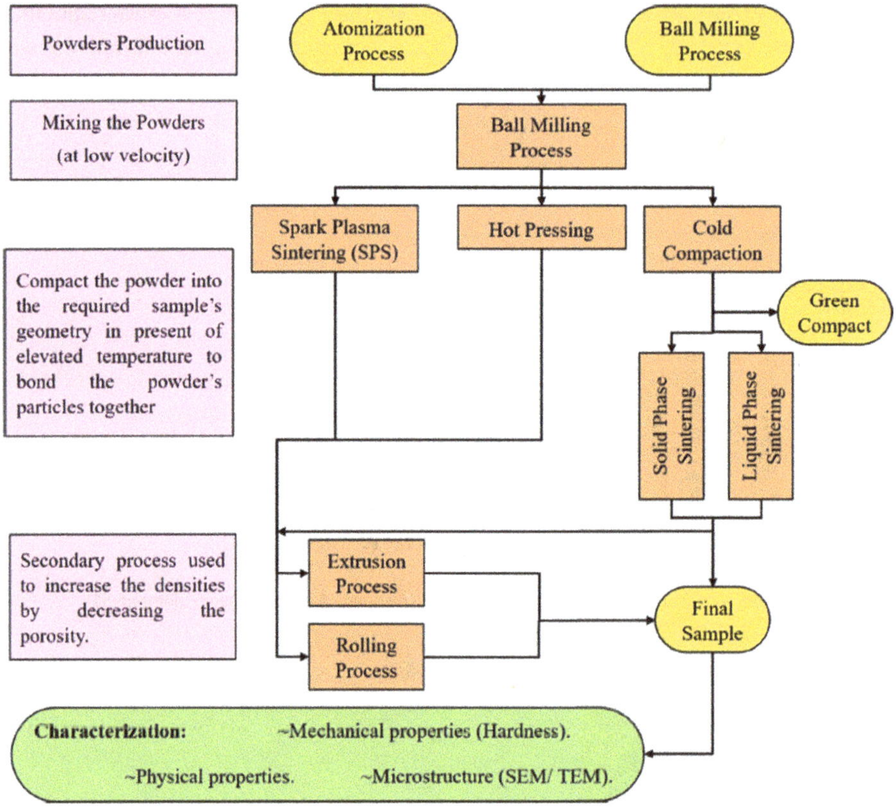

Figure 3.3: Process flow diagram for HEAs in solid-state processing route [14].

3.3.1 Powder atomization technique

Powder atomization is a process that uses a gas or a liquid stream to transform molten metallic elements into solid powder particles. The powder atomization process is shown schematically in Figure 3.4. In a two-stage atomization device, the first stage is the atomization furnace chamber, where the metallic powders, known as the charge, are melted in vacuum to the necessary temperature. The second stage involves draining of the melt liquid through a nozzle situated at the lowermost of the tundish into the atomizing chamber [14].

The atomizing nozzle splits the melt liquid stream into small droplets by a high-pressure water jet or a gas jet, which subsequently hardens instantaneously (Figure 3.4), to produce solid powder particles as the molten melt continues to flow within the tundish by gravity. The particle size of the metal powders is influenced by the "coolant pressure, coolant velocity, coolant distance, coolant flow rate, liquid mass flow rate, liquid stream velocity, impingement angle, super-heat, surface tension on metal, and the range of metal meting" [14]. Table 3.1 lists the parameters to be used during the powder manufacturing process.

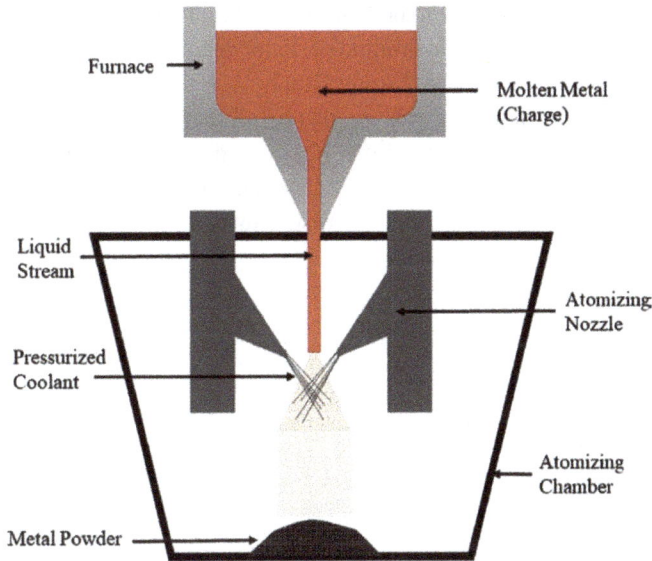

Figure 3.4: Schematic of the powder atomization process [14].

Table 3.1: Recommended process variables range for powder atomization process [14, 15].

Process variable	Range
Gas pressure (P)	1.4–42 MPa
Gas velocity (V_g)	50–150 m/s
Superheat	373–473 K
Gas flow rate	21–98 m³/h
Impingement angle	15–90°

3.3.2 Mechanical alloying

"Mechanical alloying is a non-equilibrium, high-energy powder metallurgy solid-state manufacturing method that alloys powder particles by repeatedly fracturing and cold welding" [16–18]. The processing of HEAs via the solid-state processing route, which involves mainly MA of the powder elements, followed by sintering, is covered in only about 5–10% of the research papers on HEAs, until now. Mechanical alloying is a high-energy ball milling procedure for elemental powder blends that involves elemental diffusion to produce a homogenous alloy [18].

Benjamin and his colleagues first developed this process as part of an effort to make reinforced Ni-based superalloys by oxide dispersion [19]. Fecht and colleagues published the first comprehensive report on the manufacture of nanocrystalline metals using HEBM in 1990 [20]. Figure 3.5 shows the ball used to collide elemental powder particles during milling, which includes continual particle "deformation, fracture, and welding, eventually leading to nanocrystallization" [18].

MA is a solid-state technique that produces amalgamated metal powders with specified microstructures by dispersing the insoluble phases and adding reactive alloying components. Mechanical alloying is a dry, HEBM method that generates amalgamated metal alloy powders with fine, controlled microstructures [21].

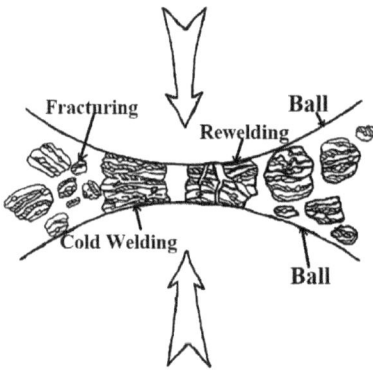

Figure 3.5: During HEBM, fracture and welding phenomena occur when ball and powder particles collide [18].

Figure 3.6 shows the scientific literature that has been published on the MA synthesis of various HEAs. Very little study has been carried out on the effects of MA and the factors that influence the HEA performance [22]. A shaker mill or a high-energy ball mill is commonly used in this procedure to compress mixed powders of known particle-sized elements in order to agglomerate them. "The initial stage deformation, a welding phase, an equiaxed particle production period, the commencement of random welding direction, and steady-state processing" are the five stages of the mechanical alloying process [23]. Composite particles form as an effect of repetitive welding and fracture of powder particles, and the structure is refined over time. MA has recently received more attention in literature than traditional processes such as melting and casting. MA assists particle size homogeneity, consolidation, and grain size reduction, as well as correct densification of the elemental powders [24]. The rate of microstructural refinement in HEAs is influenced by the input of mechanical energy along with the rate of strain hardening between the materials. Consequently, these have a favorable influence on the properties of

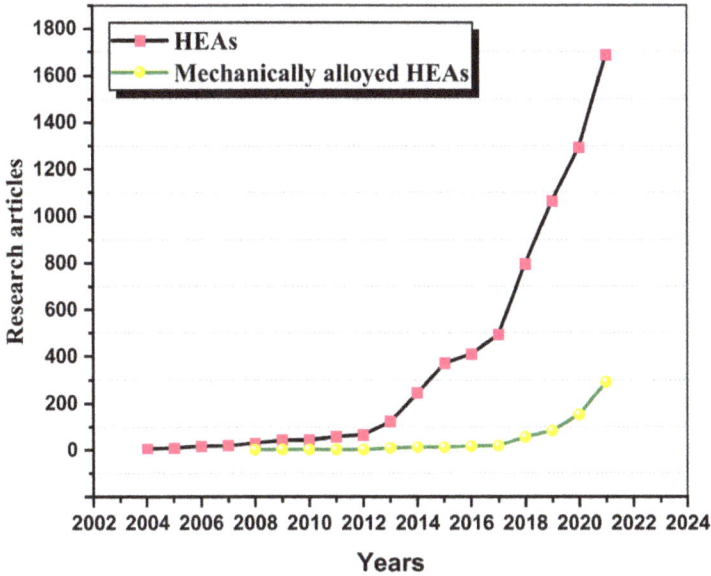

Figure 3.6: The progress of published research articles related to the development of HEAs via mechanical alloying [22].

HEAs. The MA method is thought to be a more simple and cost-effective method of producing "nanocrystalline with a homogeneous microstructure" [24]. The types of ball milling are shown in Figure 3.7. The types are tumbler, planetary, attritor-stirred, and vibratory ball mills. They are classified based on the movement type of the balls, and all are commonly made up of grinding balls in the milling chamber [25]. The nanostructured HEAs powder can be synthesized using the HEBM technique. In a **planetary ball mill**, vertical grinding jars are mounted on a horizontal disc called a sun wheel or a planetary disc, as shown in Figure 3.7. The jar rotates in a direction opposite to that of the rotation of the sun wheel, with the standard sun wheel-to-jar velocity ratios of 1:1, 1:2, and 1:3 [14]. Planetary ball milling produces nanostructured, uniform-sized powder particles (nanoscale particle). The accumulation of particles due to the increased surface energy between the powder particles is one of the most prominent drawbacks of this procedure under dry milling. Fortunately, the agglomeration may currently be reduced by adding a Process Control Agent (APC) to Mechanical alloying [26] or mechanical milling. HEBM can manufacture very high-entropy alloys by mixing pure metal powders without creating any intermetallic phases. Furthermore, HEBM can be utilized to make high-melting alloys that are hard to make using the casting/liquid metallurgical technique. WMoNbZrV HEAs, with a high melting point, were synthesized by Oleszak et al. [27]. The obtained high-entropy powders (50 h) retained their 10 nm crystallite size and the BCC structure until 700 °C, ensuring high thermal stability. Several HEBM parameters determine the quantity of the solid solution solubility, crystal size reduction, alloy formation, structural alterations, and so on. Mamifar et al. [28] examined the "effect of milling time in hours on AlCrFeCoC HEA," and found that the milling time had a significant impact on the particle size. Table 3.2 shows the most often utilized HEBM parameters for making nanostructured HEAs. Unfortunately, this method is exclusively used in laboratories and for research purposes. Contamination is the most significant disadvantage of HEBM, owing to the little amount of HEA powders produced, which has a direct impact on the final quality.

Table 3.2: HEBM parameters commonly used in the synthesis of nanostructured materials [14, 17].

Ball-to-powder ratio	5:1, 10:1, 15:1, 20:1
Sun wheel speed, in rpm	50–150
Planet vial speed, in rpm	50–450
Milling time, in h	1–100
Process control agent, to avoid agglomeration	Ethanol, gasoline, steric acid, etc.
Milling forward: pause: milling reverse time	10:10:10; 15:15:15, 20:20:20
Milling medium	Tungsten carbides, hardened stainless steel
Ball diameter, in mm	6–15
Distribution of the size of balls	Uniform and non-uniform size

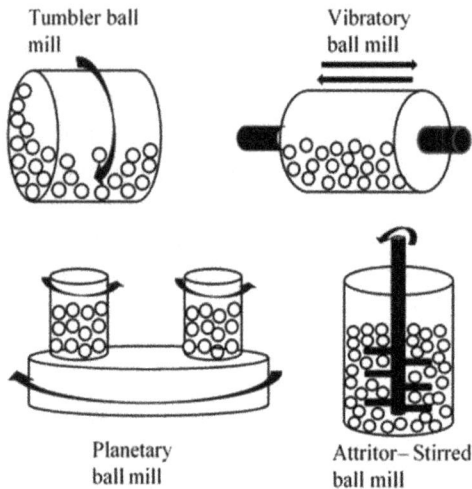

Figure 3.7: Schematic and types of ball milling [25].

3.3.3 Spark Plasma Sintering (SPS)

While compacting the milled powders, SPS utilizes the simultaneous application of high DC pulse voltage, current (800 to 1,500 A), and impact pressure (30 to100 MPa) to obtain "the dense microstructure with low grain growth in a controlled atmosphere with proper sintering parameters" [16, 29]. "During the sintering process, spark plasma and impact pressure are utilized to produce high temperatures on the particles, forcing the surface of the particles to melt" [30]. The sintering process of alloys by using SPS was divided into five stages by Chakraborty et al. – plasma generation, high heating, melting, molten particle sputtering, and neck growth [31]. As-milled powder is charged directly into a graphite-based die, which applies a uniaxial force or pressure and current at the same time to produce a completely dense alloy with superior mechanical characteristics [32]. Due to the rapid heating rate, both conductive and nonconductive particles can be densified in a short time [30]. High bulk heating and cooling rate in SPS improve its densification as well as the diffusion mechanism [32]. A faster cooling rate lowers the micro-level segregation in the milled alloy and allows for grain refinement, as seen in SPS [33]. The processing settings have a big impact on the microstructure and property of the sintered products, and because of its benefits over traditional sintering procedures, the SPS approach is frequently utilized to manufacture nanocrystalline microstructures, composite materials and advanced ceramics [34]. Not only is the sintering technique simple to use, but it also reduces grain formation in the microstructure. As a result, as compared to melting and casting, the microstructure is enhanced, resulting in superior mechanical qualities. This fabrication approach improves the properties of HEAs by strengthening the solid solution. These are some of the reasons why, in addition to its noteworthy advantages such as lesser processing time [34], flexible sintering temperature, porosity avoidance, control on various sintering parameters, and effective energy-efficient process, many researchers are increasingly favoring SPS. The SPS schematic is shown in the Figure 3.8.

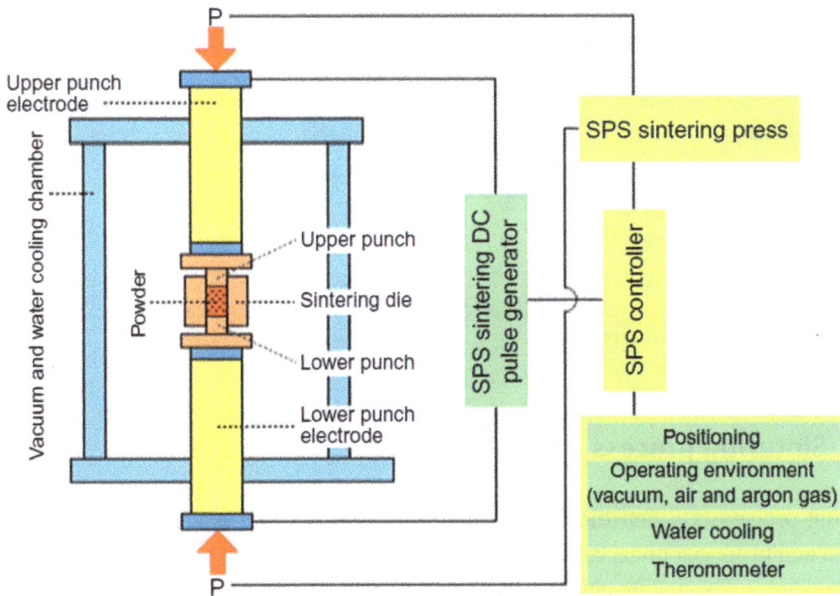

Figure 3.8: Schematic of the SPS process [18].

3.3.4 Cold uniaxial pressing process

A cold pressing (CP) method involves the manufacturing of powders as well as the consolidation of metal powders by pressing metal powder particles into a single piece, known as a green compact, using a pressing instrument at room temperature [14] The produced powder elements are to be loaded into the die (typically made of H13 steel, tool steel, EN steel) that is normally inserted between two punches during cold uniaxial pressing, shown in Figure 3.9. The punches can begin to press the metal powder particles by a process known as uniaxial compaction, because the stress is applied only in one direction. As it is still employed in mass production, the CP method is easy and cost-effective. Yuhu et al. [35] used a uniaxial pressing technique (compressed at 310 MPa) to consolidate the AlNiCrFexMo0.2CoCu HEA powder and accomplished a relative density (above 91%) after sintering. For the AlCoCrFeNi HEA powder consolidation, Shivam et al. used the CP technique (625 MPa) [36]. When this procedure was compared to the arc melting process, there was no porosity. The processing route significantly influences the microstructure and properties of HEAs [14].

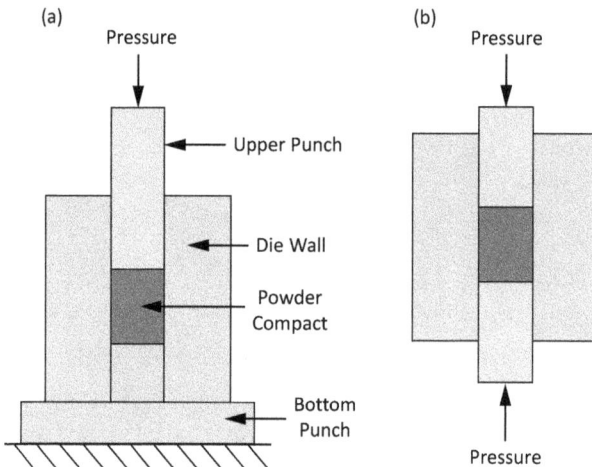

Figure 3.9: Schematic of the cold pressing (CP) process (a) single action die compaction, (b) double action die compaction [37].

3.3.5 Sintering process

Sintering process is a densification technique in which metal powder particles are crushed and produced under an inert gas atmosphere by supplying heat below the liquification point or the melting point. The goal of this technique is to connect all of the powders together in order to enhance the mechanical and physical qualities. Figure 3.10

provides a schematic of the various steps of the sintering process. Sintering is of two types – solid-state sintering and liquid-phase sintering. Liquid-phase sintering is widely utilized in difficult-to-sinter materials. Particle-to-particle diffusion bonding is possible in solid-phase sintering. The most prevalent consolidation processes for HEA powders are Spark Plasma Sintering and Vacuum Hot Pressing (VHP) sintering [14].

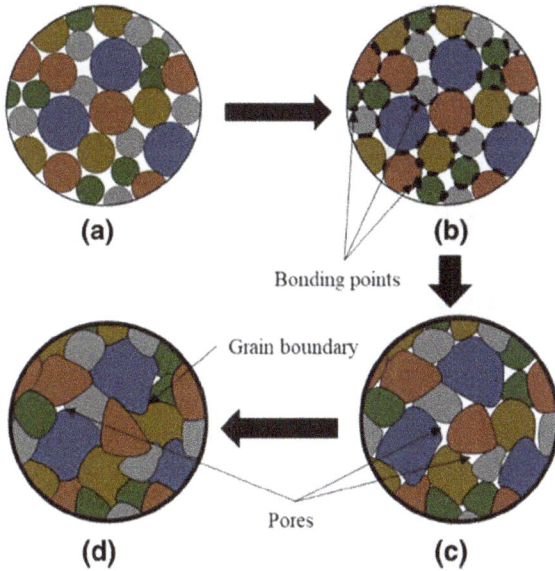

Figure 3.10: The schematic of various sintering stages:(a) Bonding points where particles are joined together in a green compact, (b) necks generated by bonding points, (c) the size of the pores within grain boundaries decreases and (d) between particles, the grain boundaries grow [16].

3.3.6 Hot pressing process

The hot pressing (HP) procedure is a simultaneous execution of both the CP and the sintering processes. The benefit of this technique is that it increases the product's density while eliminating porosities [14]. The heated powders inside the metallic die are subjected to an axial load in the HP process, shown in Figure 3.11. This HP can take place in a vacuum or at ambient temperature. HEA powders can be consolidated into bulk form using HP [38]. Mechanical properties at high temperatures are quite important in structural applications. As a result, this HEA can replace the familiar super alloys (K2135, GH1040, and A286,). The disadvantages of this HP technique are the production of intermetallic phase and the longer consolidation time. Also, in the advanced HP process, an "isostatic pressure" is used, which is known as the "hot isostatic pressing (HIP)," shown in Figure 3.12. The HIP method involves placing the materials to be combined in a metal canister (often made of steel); placing the canister in a controlled heated chamber; and then pressing the materials with the isostatic pressure (in a 360-degree direction) over an inert medium. As a result, high-density materials can be obtained.

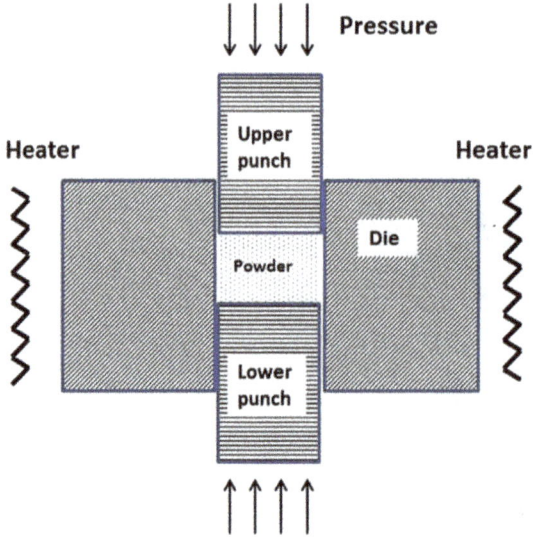

Figure 3.11: Schematic diagram of hot pressing (HP) process [39].

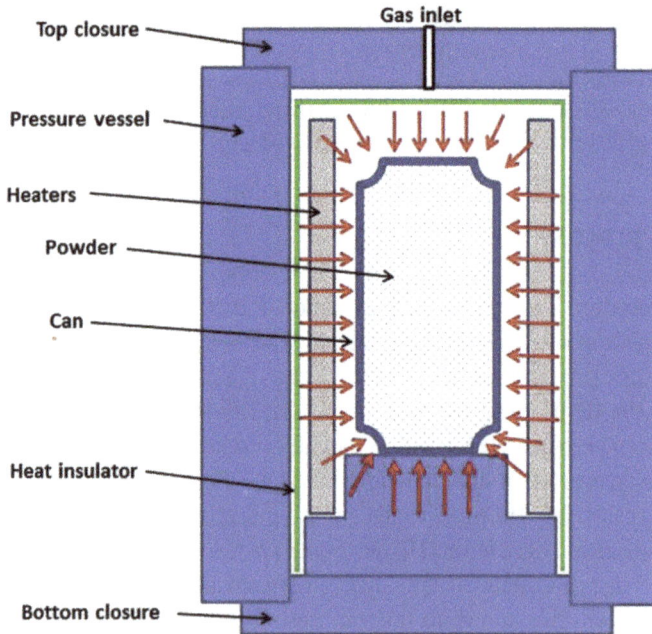

Figure 3.12: Process diagram of hot isostatic pressing (HIP) process [39].

References

[1]   J. M. Torralba, P. Alvaredo, and A. García-Junceda, "High-entropy alloys fabricated via powder metallurgy. A critical review," *Powder Metallurgy*, vol. 62, no. 2, pp. 84–114, Mar. 15. 2019, doi: 10.1080/00325899.2019.1584454.

[2]   S. Varalakshmi, M. Kamaraj, and B. S. Murty, "Synthesis and characterization of nanocrystalline AlFeTiCrZnCu high entropy solid solution by mechanical alloying," *Journal of Alloys and Compounds*, vol. 460, no. 1–2, pp. 253–257, Jul. 2008, doi: 10.1016/j.jallcom.2007.05.104.

[3]   K. B. Zhang, et al., "Nanocrystalline CoCrFeNiCuAl high-entropy solid solution synthesized by mechanical alloying," *Journal of Alloys and Compounds*, vol. 485, no. 1–2, pp. L31–L34, Oct. 2009, doi: 10.1016/j.jallcom.2009.05.144.

[4]   Y.-L. Chen, et al., "Alloying behavior of binary to octonary alloys based on Cu–Ni–Al–Co–Cr–Fe–Ti–Mo during mechanical alloying," *Journal of Alloys and Compounds*, vol. 477, no. 1–2, pp. 696–705, May. 2009, doi: 10.1016/j.jallcom.2008.10.111.

[5]   A. Kumar and M. Gupta, "An insight into evolution of light weight high entropy alloys: A review," *Metals (Basel)*, vol. 6, no. 9, p. 199, Aug. 2016, doi: 10.3390/met6090199.

[6]   O. N. Senkov, G. B. Wilks, D. B. Miracle, C. P. Chuang, and P. K. Liaw, "Refractory high-entropy alloys," *Intermetallics (Barking)*, vol. 18, no. 9, pp. 1758–1765, Sep. 2010, doi: 10.1016/j.intermet.2010.05.014.

[7]   B. Ren Ke, et al., "Powder metallurgy of high-entropy alloys and related composites: A short review," *International Journal of Minerals, Metallurgy and Materials*, vol. 28, no. 6, pp. 931–943, Jun. 01. 2021, doi: 10.1007/s12613-020-2221-y.

[8]   R. Vintila, A. Charest, R. A. L. Drew, and M. Brochu, "Synthesis and consolidation via spark plasma sintering of nanostructured Al-5356/B4C composite," *Materials Science and Engineering: A*, vol. 528, no. 13–14, pp. 4395–4407, May. 2011, doi: 10.1016/j.msea.2011.02.079.

[9]   Y. Zhang, et al., "Microstructures and properties of high-entropy alloys," *Progress in Materials Science*, vol. 61, pp. 1–93, Apr. 2014, doi: 10.1016/j.pmatsci.2013.10.001.

[10] W. Ji, et al., "Alloying behavior and novel properties of CoCrFeNiMn high-entropy alloy fabricated by mechanical alloying and spark plasma sintering," *Intermetallics (Barking)*, vol. 56, pp. 24–27, Jan. 2015, doi: 10.1016/j.intermet.2014.08.008.

[11] J. W. Yeh, "Recent progress in high-entropy alloys," *Annales de Chimie: Science Des Materiaux*, vol. 31, no. 6, pp. 633–648, Nov. 2006, doi: 10.3166/acsm.31.633-648.

[12] J.-W. Yeh, "Alloy design strategies and future trends in high-entropy alloys," *JOM*, vol. 65, no. 12, pp. 1759–1771, Dec. 2013, doi: 10.1007/s11837-013-0761-6.

[13] J.-W. Yeh, et al., "Nanostructured high-entropy alloys with multiple principal elements: Novel alloy design concepts and outcomes," *Advanced Engineering Materials*, vol. 6, no. 5, pp. 299–303, May. 2004, doi: 10.1002/adem.200300567.

[14] Y. A. Alshataif, S. Sivasankaran, F. A. Al-Mufadi, A. S. Alaboodi, and H. R. Ammar, "Manufacturing methods, microstructural and mechanical properties evolutions of high-entropy alloys: A review," *Metals and Materials International*, vol. 26, no. 8, pp. 1099–1133, Aug. 2020, doi: 10.1007/s12540-019-00565-z.

[15] A. Lawley, "Atomization of specialty alloy powders," *JOM*, vol. 33, no. 1, pp. 13–18, Jan. 1981, doi: 10.1007/BF03354395.

[16] S. Yadav, K. Biswas, and A. Kumar, "Spark plasma sintering of high entropy alloys," In: *Spark Plasma Sintering of Materials*, Springer International Publishing, 2019, pp. 539–571. doi: 10.1007/978-3-030-05327-7_19.

[17] C. Suryanarayana, "Mechanical alloying and milling," *Progress in Materials Science*, vol. 46, no. 1–2, pp. 1–184, Jan. 2001, doi: 10.1016/S0079-6425(99)00010-9.

[18] B. S. Murty, J. W. Yeh, and S. Ranganathan, "Synthesis and processing," In: *High Entropy Alloys*, Elsevier, 2014, pp. 77–89. doi: 10.1016/b978-0-12-800251-3.00005-5.

[19] J. S. Benjamin, "Dispersion strengthened superalloys by mechanical alloying," *Metallurgical and Materials Transactions B*, vol. 1, no. 10, pp. 2943–2951, Oct. 1970, doi: 10.1007/BF03037835.

[20] H. J. Fecht, E. Hellstern, Z. Fu, and W. L. Johnson, "Nanocrystalline metals prepared by high-energy ball milling," *Metallurgical Transactions A*, vol. 21, no. 9, pp. 2333–2337, Sep. 1990, doi: 10.1007/BF02646980.

[21] Y., G. M. C., Y. J. W., L. P. K., and Z. Y. Zhang. *High-Entropy Alloys*. Cham: Springer International Publishing, 2016. doi: 10.1007/978-3-319-27013-5.

[22] A. Kumar, A. Singh, and A. Suhane, "A critical review on mechanically alloyed high entropy alloys: Processing challenges and properties," *Materials Research Express*, vol. 9, no. 5, 2022, doi: 10.1088/2053-1591/ac69b3.

[23] G. Mucsi, "A review on mechanical activation and mechanical alloying in stirred media mill," *Chemical Engineering Research & Design*, vol. 148, pp. 460–474, Aug. 2019, doi: 10.1016/j.cherd.2019.06.029.

[24] M. Hebda, S. Gądek, M. Skałoń, and J. Kazior, "Effect of mechanical alloying and annealing on the sintering behaviour of AstaloyCrL powders with SiC and carbon addition," *Journal of Thermal Analysis and Calorimetry*, vol. 113, no. 1, pp. 395–403, Jul. 2013, doi: 10.1007/s10973-013-3218-9.

[25] Y. W. Sitotaw, N. G. Habtu, A. Y. Gebreyohannes, S. P. Nunes, and T. van Gerven, "Ball milling as an important pretreatment technique in lignocellulose biorefineries: A review," *Biomass Convers Biorefin*, Aug. 2021, doi: 10.1007/s13399-021-01800-7.

[26] R. Clinktan, V. Senthil, K. R. Ramkumar, S. Sivasankaran, and F. A. Al-Mufadi, "Effect of boron carbide nano particles in CuSi4Zn14 silicone bronze nanocomposites on matrix powder surface morphology and structural evolution via mechanical alloying," *Ceramics International*, vol. 45, no. 3, pp. 3492–3501, Feb. 2019, doi: 10.1016/j.ceramint.2018.11.007.

[27] D. Oleszak, A. Antolak-Dudka, and T. Kulik, "High entropy multicomponent WMoNbZrV alloy processed by mechanical alloying," *Materials Letters*, vol. 232, pp. 160–162, Dec. 2018, doi: 10.1016/j.matlet.2018.08.060.

[28] A. Emamifar, B. Sadeghi, P. Cavaliere, and H. Ziaei, "Microstructural evolution and mechanical properties of AlCrFeNiCoC high entropy alloy produced via spark plasma sintering," *Powder Metallurgy*, vol. 62, no. 1, pp. 61–70, Jan. 2019, doi: 10.1080/00325899.2019.1576389.

[29] W. Ji, et al., "Alloying behavior and novel properties of CoCrFeNiMn high-entropy alloy fabricated by mechanical alloying and spark plasma sintering," *Intermetallics (Barking)*, vol. 56, pp. 24–27, Jan. 2015, doi: 10.1016/j.intermet.2014.08.008.

[30] P. F. Yu, et al., "The high-entropy alloys with high hardness and soft magnetic property prepared by mechanical alloying and high-pressure sintering," *Intermetallics (Barking)*, vol. 70, pp. 82–87, Mar. 2016, doi: 10.1016/j.intermet.2015.11.005.

[31] B. Wang, et al., "Mechanical properties and microstructure of the CoCrFeMnNi high entropy alloy under high strain rate compression," *Journal of Materials Engineering and Performance*, vol. 25, no. 7, pp. 2985–2992, Jul. 2016, doi: 10.1007/s11665-016-2105-5.

[32] B. Liu, et al., "Microstructure and mechanical properties of equimolar FeCoCrNi high entropy alloy prepared via powder extrusion," *Intermetallics (Barking)*, vol. 75, pp. 25–30, Aug. 2016, doi: 10.1016/j.intermet.2016.05.006.

[33] J. Wang, et al., "Flow behavior and microstructures of powder metallurgical CrFeCoNiMo0.2 high entropy alloy during high temperature deformation," *Materials Science and Engineering: A*, vol. 689, pp. 233–242, Mar. 2017, doi: 10.1016/j.msea.2017.02.064.

[34] Ł. Rogal, D. Kalita, and L. Litynska-Dobrzynska, "CoCrFeMnNi high entropy alloy matrix nanocomposite with addition of Al 2 O 3," *Intermetallics (Barking)*, vol. 86, pp. 104–109, Jul. 2017, doi: 10.1016/j.intermet.2017.03.019.

[35] F. Yuhu, Z. Yunpeng, G. Hongyan, S. Huimin, and H. Li, "AlNiCrFexMo0.2CoCu high entropy alloys prepared by powder metallurgy," *Rare Metal Materials and Engineering*, vol. 42, no. 6, pp. 1127–1129, Jun. 2013, doi: 10.1016/S1875-5372(13)60074-0.

[36] V. Shivam, J. Basu, V. K. Pandey, Y. Shadangi, and N. K. Mukhopadhyay, "Alloying behaviour, thermal stability and phase evolution in quinary AlCoCrFeNi high entropy alloy," *Advanced Powder Technology*, vol. 29, no. 9, pp. 2221–2230, Sep. 2018, doi: 10.1016/j.apt.2018.06.006.

[37] W. Wang, H. Qi, P. Liu, Y. Zhao, and H. Chang, "Numerical simulation of densification of Cu–Al mixed metal powder during axial compaction," *Metals (Basel)*, vol. 8, no. 7, p. 537, Jul. 2018, doi: 10.3390/met8070537.

[38] C. Sun, P. Li, S. Xi, Y. Zhou, S. Li, and X. Yang, "A new type of high entropy alloy composite Fe18Ni23Co25Cr21Mo8WNb3C2 prepared by mechanical alloying and hot pressing sintering," *Materials Science and Engineering: A*, vol. 728, pp. 144–150, Jun. 2018, doi: 10.1016/j.msea.2018.05.022.

[39] J. M. Torralba, "Improvement of mechanical and physical properties in powder metallurgy," In: *Comprehensive Materials Processing*, Elsevier, 2014, pp. 281–294. doi: 10.1016/B978-0-08-096532-1.00316-2.

Rituraj Chandrakar, Om Prakash, Rajesh Kumar,
Hanuman Reddy Tiyyagura, Saurabh Chandraker

Chapter 4
Melting and casting route

Abstract: The melting and casting route is the most common and relatively cheap route of production of high-entropy alloys. In this route, the constituent elements are mixed in liquid state. Multicomponent alloys in the shape of buttons, rods, ribbons, and bars have been created using the melting and casting route, with various cooling rates. Vacuum arc melting is the most common melting and processing process. This chapter reviews melting and casting routes and related synthesis techniques in manufacturing of high-entropy alloys.

Keywords: High-entropy alloys (HEAs), Liquid melting route, Vacuum arc melting (VAM), Vacuum induction melting (VIM), Bridgman solidification technique (BST), Laser cladding (LC), Laser-Enhanced Net Shape (LENS)

4.1 Introduction

The melting and casting route is the most common and relatively cheap fabrication route. The melting and casting processes, however, have a significant tendency for a heterogeneous structure with the element's segregation and flaws [1]. Multicomponent alloys in the shape of button, rods, ribbons, and bars have been created using melting and casting procedures with equilibrium and non-equilibrium cooling rates. Vacuum arc melting (VAM) is the most common melting and processing process [2]. Melting and solidification techniques are used to manufacture bulk high-entropy alloys (HEAs) from a liquid state. The pure solid metals are melted into a liquid condition and homogeneously combined throughout the melting. The liquid metal combination is then solidified to generate bulk HEAs under a different cooling rate [3].

Rituraj Chandrakar, Saurabh Chandraker, Department of Mechanical Engineering, NMDC DAV Polytechnic, Geedam, Dantewada, Chhattisgarh 49441, India, e-mail: rituraj.chandraker@gmail.com And; Department of Mechanical Engineering, National Institute of Technology Karnataka (NITK), Surathkal, Mangalore 575025, India
Om Prakash, Department of Mechanical Engineering, Jhada Sirha Government Engineering College, Jagdalpur, Bastar, Chhattisgarh 494001, India
Rajesh Kumar, Department of Mechanical Engineering, CSIT, Durg, Chhattisgarh 491001, India
Hanuman Reddy Tiyyagura, Rudolfovo Science and Technology, Centre Novo mesto Podbreznik 15, 8000 Novo mesto, Slovenia

https://doi.org/10.1515/9783110769470-004

4.2 Melting and casting route

Melting and casting techniques can also be used to make HEAs. VAM is mostly used in this route. The evaporation of low melting materials is the fundamental disadvantage of liquid-state processing. Resistance and vacuum induction heating furnaces can be used to manage it. Another problem of the casting technique for HEAs is the development of "**heterogeneous microstructure**," due to the slower rate of solidification [4]. Melting and casting is the most extensively used method for producing HEAs. Figure 4.1 shows the number of studies on HEAs that have been published and organized by the synthesis route. Figure 4.1 clearly shows that the casting approach leads the processing routes in bulk, with this route producing about 75% of the HEA articles published so far. VAM has been used to manufacture the most popular of HEAs that have been documented so far, with some vacuum induction melting (VIM) cases as well [2].

Melting and casting route includes arc melting, VIM LC method, and LENS (Laser-Enhanced Net Shape) [2].

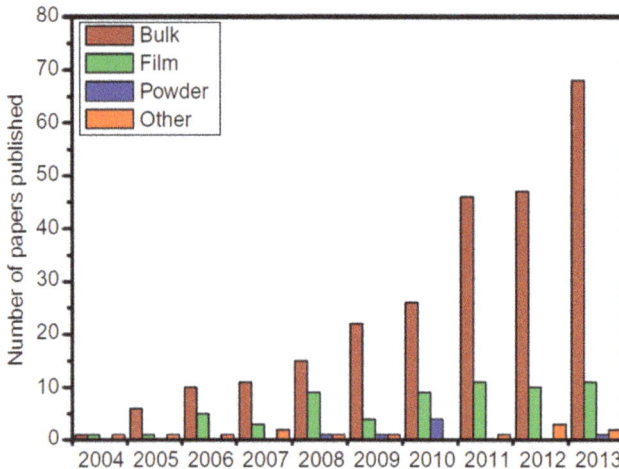

Figure 4.1: A number of publications on HEAs that have been published as a result of various processing methods [2].

4.2.1 Arc melting

The arc melting technique is the most extensively used processing technique for the fabrication of HEAs. Figure 4.2 shows a typical arc melting technique. The electrode temperature of an arc melting furnace can reach above 3,000 °C, which is hot enough to melt most elements used in HEAs and can be adjusted by altering the electrical power. As a result, the arc melting technique is advantageous for homogeneous melt mixing of high melting point materials, such as refractory elements. The **heterogeneous microstructure** formed by several segregation mechanisms due to slower rate of solidification is one of the major disadvantages in this route. **Dendritic** (DR) with **Interdendritic** (ID) segregation characteristics are mostly seen in the microstructure of HEAs that are normally synthesized by the arc melting method [2]. However, the temperature gradient from surface to center during solidification causes heterogeneous microstructure (dendritic with interdendritic segregation) in the final HEAs, making the microstructure and characteristics of as-cast HEAs less controlled. Furthermore, in refractory HEAs cast from liquid state, substantial discrepancies in elemental melting temperatures can result in elemental segregation [5, 6]. As the arc melting approach limits the as-cast alloy shapes to rods and buttons, complicated shapes for advanced applications are difficult to cast. However, the drawback of this method is that certain low boiling point metals may evaporate during alloy formation. Resistance and vacuum induction heating methods have been used to create the alloys, in these circumstances [2].

The arc melting technique involves placing metal ingots or powders on crucibles and melting them with a tungsten electric arc that strikes in an inert gas (argon gas), after vacuuming the chamber to prevent oxidation. The metal ingots or powders were repeatedly melted by the arc melting process and solidification by the coolant under the metallic crucible, to ensure that the alloy was homogeneous. All the constituent metals are fully combined in the liquid stage, before being solidified in the crucible during arc melting. To achieve chemical homogeneity of the alloys, several recurrent

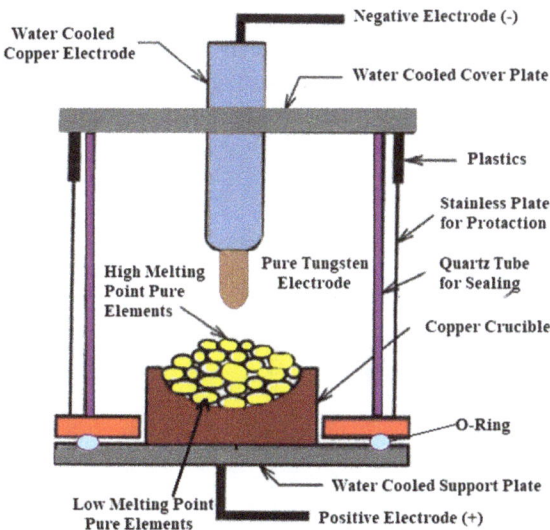

Figure 4.2: Arc melting process [7].

melting and solidification processes are frequently performed. In the bowl-like cruci-ble, solidified ingots have a button-like shape, and rod-shaped ingots can be made with a faster cooling rate during solidification. Additionally, machined specimens for tests like compression and tensility can be made from the samples. A range of HEAs with excellent properties have been synthesized using this casting approach. For ex-ample, Zhou et al. [8] used this process to make AlCoCrFeNiTi$_x$ (x in molar ratio) HEAs with BCC single solid-solution and exceptional compressive attributes such as ultra-high fracture strength and considerable strain-hardening capability. By significantly raising the cooling rate during solidification and reducing the diameter of cast prod-ucts, the alloy mechanical strength and flexibility of HEAs increased [9].

4.2.2 Vacuum induction melting

It is a method of heating electrically conducting materials by inducing a magnetic field via electromagnetic induction. The electrical current, as it passes through the conducting materials, results in eddy current, which is subsequently employed to heat the materials. As shown in Figure 4.3, the ingot inside the furnace blocks the input current, causing a magnetic field to form through the electrically conducting materials, causing them to melt quickly from within. Then, to obtain homogeneous HEA, the ingot is chilled and remelted, numerous times [10–12].The molten condition of HEAs is to be put into the desired mold and allowed to solidify. The melting of HEAs must take place in a high-vacuum environment. The melted HEAs were poured into a copper mold and cooled to room temperature, before being used. The casting was done in an argon-free environment. In comparison to the arc melting process, the VIM method has the benefit of accurately controlling the heating and cooling speeds

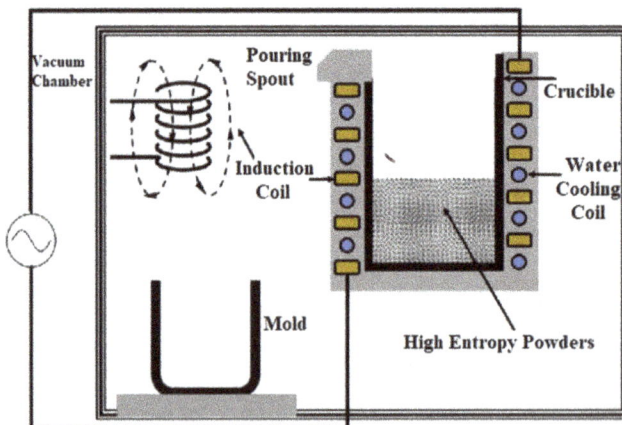

Figure 4.3: Process flow diagram for VIM [4].

as well as achieving homogeneity. Furthermore, the vacuum environment avoids oxidation. The fundamental constraint of this approach is the low surface finish, which necessitates the use of a secondary machining procedure [4].

4.2.3 Bridgman solidification technique

Compared to conventional casting, the Bridgman solidification technique (BST) can be used for better control of the microstructure and optimization of the characteristics of HEAs. For rod and cylindrical-shaped samples obtained by BST, thermal conduction and extraction are heavily concentrated along the longitudinal direction, thus ensuring the direction of microstructural growth. Two critical synthesis parameters, specifically the temperature gradient (G) and growth rate, can be controlled by regulating the power of heating and withdrawal velocities, thus ensuring microstructural morphology and size. Finally, the two elements produce HEAs with tunable microstructures and characteristics [13].The BST processing and sample location are depicted schematically in Figure 4.4. The target alloys are initially cast using a normal casting process into certain rod-shaped

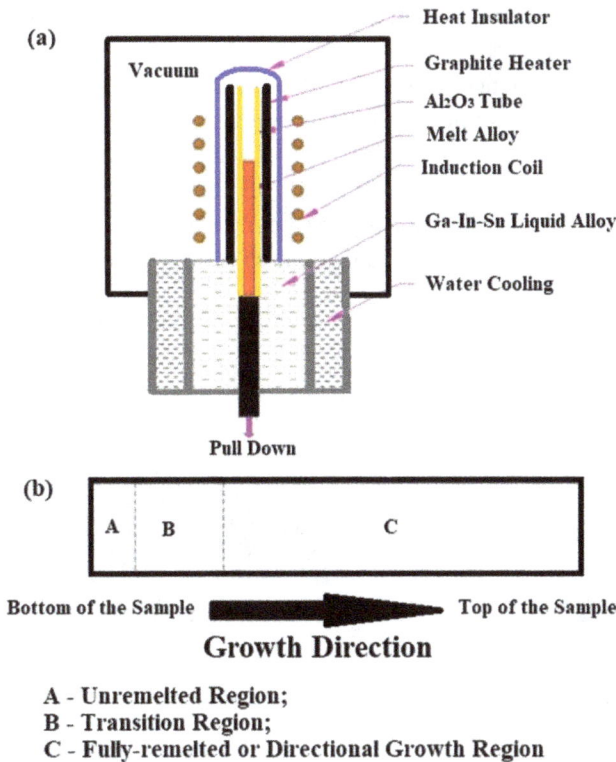

Figure 4.4: (a) Schematic of the BST and (b) the sample location [14].

samples. The synthesized samples are then broken up into little pieces and put inside a tube made of alumina. The tube is then inductively heated to a completely melted state by adjusting the power of heating and proper holding time. The BST process is then completed by applying a temperature gradient of 40–45 K/mm and a withdrawal velocity (R) of 5–2,000 m/s. Figure 4.4 (b) shows the sample location – section A, the unremelted region; section B, the transition region; and section C, the fully remelted region – are formed as a result of the liquid alloys being directed into the "Ga-In-Sn liquid alloy" surrounded by water and cooled [14].

4.2.4 Laser cladding method

A quick prototype technique called "laser cladding" (LC) combines "laser technology, with computer-aided manufacturing, and the advanced control system" [15]. The primary fabrication method for HEAs-based coatings is LC, where the substrate surface layer and the pre-placed powders or wires are simultaneously melted and rapidly consolidated by the laser's heat. In the hard facing method known as LC, the coating material and a small layer of the "substrate" are melted together by a powerful laser beam. This results in a coating that is 2–50 mm thick, pore- and crack-free, has low porosity, and is perfectly attached to the substrate [16]. When compared to laser surface alloying, sound metallurgical connection between the substrate and the cladded coatings can be accomplished, and the coating dilution is minimal [17]. Hardness and surface-dependent qualities including wear, corrosion, oxidation, and, to a lesser extent, fatigue resistance, may be improved through the process [16]. Due to the extremely high heating temperature and rapid cooling rate achieved during LC, the microstructure of the coatings is mostly fine and homogenous, even with the development of

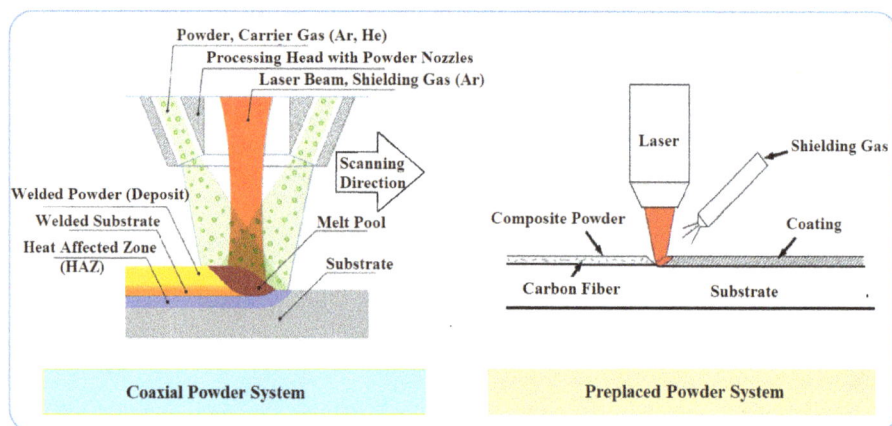

Figure 4.5: Schematic diagram of a coaxial powder system and a pre-placed powder system [18, 19].

"amorphous and nanocrystalline materials." However, the inherent properties of HEAs simplify and partially constrain the phase compositions of the deposited layer of coating materials, which are restricted to intermetallic compounds and single-phase solid solutions. These might help explain why HEA-based laser cladded coatings have excellent mechanical and functional characteristics [17]. It can be further divided into four varieties, based on the method used to feed the powder: coaxial powder systems, pre-placed powder systems, off-axis powder systems, and wire feeding systems. Coaxial powder systems and pre-placed powder systems are often the most frequently employed LC techniques. A schematic diagram of a coaxial powder system and a pre-placed powder system is shown in Figure 4.5 [18, 19].

4.2.5 Laser-Enhanced Net Shape (LENS)

Another well-known method of additive manufacturing that may create net-shaped metallic objects with multipart geometries without the need for a powder bed is Laser Enhanced Net Shaping (LENS) [20]. Figure 4.6 displays a diagram of this method. In comparison to selective laser melting, which uses a powder bed, LENS technology creates the component by supplying a powder through a narrow nozzle and irradiating a powerful laser beam with extremely high energy to melt and keep depositing in a layer-by-layer pattern over a build base-plate. The build platform descends gradually with the deposition of each layer, and until the required component is realized, this process is repeated. Although the LENS technology was primarily created to make complicated geometrical components, it also has a place in the repair and refurbishment of harmed buildings and components. LENS has a limited number of issues,

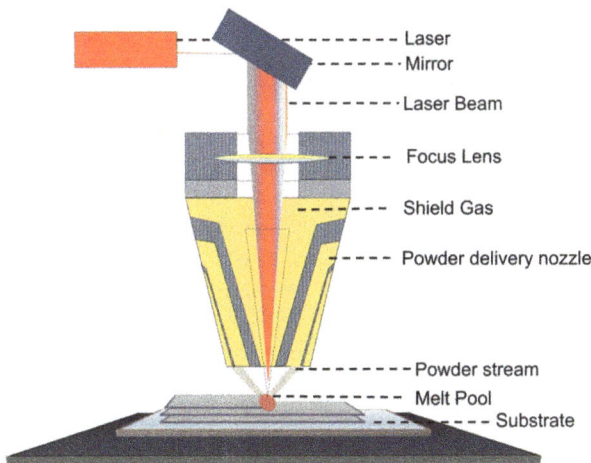

Figure 4.6: Schematic illustration of the LENS process [22].

including the necessity for post-processing, inferior component surface polish, and component deformation brought on by residual pressures [21].

References

[1]   L. J. Santodonato, et al., "Deviation from high-entropy configurations in the atomic distributions of a multi-principal-element alloy," *Nature Communications*, vol. 6, no. 1, p. 5964, May. 2015, doi: 10.1038/ncomms6964.

[2]   B. S. Murty, J. W. Yeh, and S. Ranganathan, "Synthesis and processing," In: *High Entropy Alloys*, Elsevier, 2014, pp. 77–89. doi: 10.1016/b978-0-12-800251-3.00005-5.

[3]   Y. Zhang and Q. Xing, "High entropy alloys: Manufacturing routes," In: *Encyclopedia of Materials: Metals and Alloys*, Elsevier, 2021, pp. 327–338. doi: 10.1016/B978-0-12-803581-8.12123-X.

[4]   Y. A. Alshataif, S. Sivasankaran, F. A. Al-Mufadi, A. S. Alaboodi, and H. R. Ammar, "Manufacturing methods, microstructural and mechanical properties evolutions of high-entropy alloys: A review," *Metals and Materials International*, vol. 26, no. 8, pp. 1099–1133, Aug. 2020, doi: 10.1007/s12540-019-00565-z.

[5]   O. N. Senkov, G. B. Wilks, D. B. Miracle, C. P. Chuang, and P. K. Liaw, "Refractory high-entropy alloys," *Intermetallics (Barking)*, vol. 18, no. 9, pp. 1758–1765, Sep. 2010, doi: 10.1016/j.intermet.2010.05.014.

[6]   J. P. Couziné, et al., "Microstructure of a near-equimolar refractory high-entropy alloy," *Materials Letters*, vol. 126, pp. 285–287, Jul. 2014, doi: 10.1016/j.matlet.2014.04.062.

[7]   Y. Y. Chen, T. Duval, U. D. Hung, J. W. Yeh, and H. C. Shih, "Microstructure and electrochemical properties of high entropy alloys – A comparison with type-304 stainless steel," *Corrosion Science*, vol. 47, no. 9, pp. 2257–2279, Sep. 2005, doi: 10.1016/j.corsci.2004.11.008.

[8]   Y. J. Zhou, Y. Zhang, Y. L. Wang, and G. L. Chen, "Solid solution alloys of AlCoCrFeNiTix with excellent room-temperature mechanical properties," *Applied Physics Letters*, vol. 90, no. 18, p. 181904, Apr. 2007, doi: 10.1063/1.2734517.

[9]   F. J. Wang, Y. Zhang, G. L. Chen, and H. A. Davies, "Cooling rate and size effect on the microstructure and mechanical properties of AlCoCrFeNi high entropy alloy," *Journal of Engineering Materials and Technology*, vol. 131, no. 3, Jul. 2009, doi: 10.1115/1.3120387.

[10] K. V. Yusenko, et al., "High-pressure high-temperature tailoring of high entropy alloys for extreme environments," *Journal of Alloys and Compounds*, vol. 738, pp. 491–500, Mar. 2018, doi: 10.1016/j.jallcom.2017.12.216.

[11] Y. Liu, et al., "Microstructure and mechanical properties of refractory HfMo0.5NbTiV0.5Six high-entropy composites," *Journal of Alloys and Compounds*, vol. 694, pp. 869–876, Feb. 2017, doi: 10.1016/j.jallcom.2016.10.014.

[12] Y. Du, Y. Lu, T. Wang, T. Li, and G. Zhang, "Effect of electromagnetic stirring on microstructure and properties of Al0.5CoCrCuFeNi alloy," *Procedia Engineering*, vol. 27, pp. 1129–1134, 2012, doi: 10.1016/j.proeng.2011.12.562.

[13] Y., G. M. C., Y. J. W., L. P. K., and Z. Y. Zhang, *High-Entropy Alloys*. Cham: Springer International Publishing, 2016. doi: 10.1007/978-3-319-27013-5.

[14] S. G. Ma, S. F. Zhang, M. C. Gao, P. K. Liaw, and Y. Zhang, "A successful synthesis of the CoCrFeNiAl0.3 single-crystal, high-entropy alloy by Bridgman solidification," *JOM*, vol. 65, no. 12, pp. 1751–1758, Dec. 2013, doi: 10.1007/s11837-013-0733-x.

[15] T. M. Yue, Y. P. Su, and H. O. Yang, "Laser cladding of Zr65Al7.5Ni10Cu17.5 amorphous alloy on magnesium," *Materials Letters*, vol. 61, no. 1, pp. 209–212, Jan. 2007, doi: 10.1016/j.matlet.2006.04.033.

[16] R. Vilar, "Laser cladding," *Journal of Laser Applications*, vol. 11, no. 2, pp. 64–79, Apr. 1999, doi: 10.2351/1.521888.

[17] J. Li, Y. Huang, X. Meng, and Y. Xie, "A review on high entropy alloys coatings: Fabrication processes and property assessment," *Advanced Engineering Materials*, vol. 21, no. 8, p. 1900343, Aug. 2019, doi: 10.1002/adem.201900343.

[18] F. Wirth and K. Wegener, "A physical modeling and predictive simulation of the laser cladding process," *Addition Manufacturing*, vol. 22, pp. 307–319, Aug. 2018, doi: 10.1016/j.addma.2018.05.017.

[19] J. Lei, C. Shi, S. Zhou, Z. Gu, and L.-C. Zhang, "Enhanced corrosion and wear resistance properties of carbon fiber reinforced Ni-based composite coating by laser cladding," *Surface & Coatings Technology*, vol. 334, pp. 274–285, Jan. 2018, doi: 10.1016/j.surfcoat.2017.11.051.

[20] A. Bandyopadhyay, B. V. Krishna, W. Xue, and S. Bose, "Application of Laser Engineered Net Shaping (LENS) to manufacture porous and functionally graded structures for load bearing implants," *Journal of Materials Science: Materials in Medicine*, vol. 20, no. S1, pp. 29–34, Dec. 2009, doi: 10.1007/s10856-008-3478-2.

[21] S. A. Kumar and R. V. S. Prasad, "Basic principles of additive manufacturing: Different additive manufacturing technologies," In: *Additive Manufacturing*, Elsevier, 2021, pp. 17–35. doi: 10.1016/B978-0-12-822056-6.00012-6.

[22] S. Guddati, A. S. K. Kiran, M. Leavy, and S. Ramakrishna, "Recent advancements in additive manufacturing technologies for porous material applications," *The International Journal of Advanced Manufacturing Technology*, vol. 105, no. 1–4, pp. 193–215, Nov. 2019, doi: 10.1007/s00170-019-04116-z.

Anil Kumar, Poonam Diwan, Jagesvar Verma, Arun Kumar Sao

Chapter 5
Effect of processing routes on the microstructure/phases of high-entropy alloys

Abstract: In the current work, an effort has been made to assess the impact of the processing method on the phase evolution and the mechanical characteristics of CoCrCuFe-NiSix high-entropy alloys (x = 0, 0.3, 0.6, and 0.9 atomic ratios). The CoCrCuFeNiSix high-entropy alloys were made by spark plasma sintering and vacuum arc melting. After spark plasma sintering, the X-ray diffraction data show a face-centered cubic structure and a sigma phase. It is also noted that the inclusion of Si promotes the production of the sigma phase. As opposed to this, a sample that is papered using the arc meting method displayed a face-centered cubic structure up to 0.6 Si content. Ni3Si is created when Si is increased further (by 0.6 and 0.9). The increase in Si concentration from 0 to 0.9 in both synthesis processes led to increased microhardness and enhanced the wear resistance. However, due to the production of sigma phases, the spark plasma sintered samples outperformed the arc-melted samples in terms of physical attributes.

Keywords: High-entropy alloy (HEAs), Spark plasma sintering (SPS), Arc melting (AM), X-ray diffraction (XRD)

5.1 Introduction

Due to their exceptional characteristics and distinct multi-element solid-solution architectures, high-entropy alloys have recently gained a lot of study interest [1]. Due to their high mixing entropy, these alloys frequently form a single phase or a multiphase structure. The development of **complex phases** or intermetallic compounds is suppressed by the high mixing entropy [2]. Typically, the phases produced by HEAs have a simple FCC, BCC, or a combination of the two [3]. There have been reports of "high-entropy alloys with good mechanical strength, **ductility**, high temperature

---

Anil Kumar, Department of Mechanical Engineering, Bhilai Institute of Technology, Durg, Chhattisgarh 491001, India, email: anilmech2010@gmail.com
Poonam Diwan, Department of Mechanical Engineering, Vishwavidyalaya Engineering College, Ambikapur, Surguja, Chhattisgarh 497001, India
Jagesvar Verma, Department of Manufacturing Engineering, National Institute of Advanced Manufacturing Technology, Near Kanchnatoli, Hatia, Ranchi, Jharkhand 834003, India
Arun Kumar Sao, Department of Mechanical Engineering, NMDC DAV Polytechnic College Geedam, Dantewada, Chhattisgarh 494441, India

https://doi.org/10.1515/9783110769470-005

strength, magnetic property, high corrosion resistance, good mechanical strength at cryogenic temperature, and even superconductivity" [4].

HEAs can be purchased in a variety of shapes, including as slabs, ingots, powders, thin films, or in bulk. However, casting and powder metallurgy are the two methods most frequently used to create HEAs (arc melting). Praveen et al. [5] used powder metallurgy to create the CuCrCoFeNi HEA. Two FCC and a minor amount of the other phase were reported to have formed. The same alloy was made by Tong et al. [6] using a casting method, and they noted the existence of a single phase (FCC). A study of Ji et al. [7] also looked at the phase evolution of an AlCoCrFeNi alloy that had been produced by MA, and then solidified by SPS. After SPS, they observed the development of a combination of the FCC and BCC phases. However, when the identical alloy composition was produced by Wang et al. [8] using the arc melting technique, just one BCC phase emerged. Similar to this, Qui et al. [9] used powder metallurgy to create an AlCrFeNiCoCu high-entropy alloy, and after **hydraulically pressing** the alloy, they found a combination of the BCC and FCC phases. Shaysultanov et al. [10] also used induction melting to create this alloy, and they noted three phases: BCC, FCC, and B2. As a result, it is important to look at how the processing method for a specific HEA affects the phase development. The phases in HEAs can be obtained from the multicomponent constituents through a range of transformations offered by various processing techniques. As a result, the processing approach will obviously have a substantial impact on the phase evolution in HEA.

Additionally, the majority of studies described in the literature focused on the variation in atomic proportion of a single transition element in the HEAs' constituent makeup. It is crucial to remember that the addition of "Si" always benefits both the structure and the properties of traditional alloys, like steel. On the other hand, to the author's knowledge, no one has ever looked into the impact of Si on CuCrCoFeNi. Si addition is usually beneficial to the structure, along with the improvement of mechanical characteristics in typical materials like steels. It is significant to note that a large number of earlier papers have examined the impact of elements on the phase evolution and on the enhancement of the mechanical properties of HEAs. The impact of non-metallic elements like Si, however, has only been the subject of a relatively small number of investigations. Therefore, an effort has been made in the current study to explore the impact of Si on the phase evolution and the mechanical characteristics of the CuCrCoFeNi alloy as well as the "impact of the processing route on the phase evolution and the physical characteristics of HEAs made by spark plasma sintering and arc melting" [3].

5.2 Materials and methods

By using elemental powders of Co, Cr, Cu, Fe, Ni, and Si, spark plasma sintering and vacuum arc melting were used to create the CoCrCuFeNiSix (x = 0, 0.3, 0.6, and 0.9) HEAs. Only the right mixing and blending of the constituent powders were done in a high intensity ball mill for 10 min. Then, using a press machine (Hydraulic) and a compaction die, the appropriately mixed and blended powders were successfully compacted into cylinder-shaped samples, with diameters of around 20 mm, under a uniaxial pressure of 1 GPa. Two distinct processing methods were used to produce the bulk samples. In the first method, the SPS condensed the mixed powders at 1,000 °C for 5 min under a constant pressure of 60 MPa in a die (made by graphite) with 10 mm inner diameter. Temperatures between 570 °C and 800 °C, with a heating rate of 100 °C per min was employed, which was then lowered to 50 °C per min, till the temperature reached 1,000 °C. The formed green compacts were get-melted in the second method in an argon (Ar) environment using a water-cooled copper mold. To increase the chemical homogeneity, the obtained samples were melted at least six times before being eventually cast.

5.3 Results and discussion

CoCrCuFeNiSi$_x$ HEAs, with Si addition (x = 0, 0.3, 0.6, and 0.9), produced by SPS have an X-ray diffraction pattern that shows the production of F1 and F2 (FCC phases) as well as the sigma phase (Figure 5.1(a)). Due to the limited solubility of Cu with the other elements in the system, Cu has a high positive ΔH_{mix} (mixing enthalpy). Cu, therefore, has a tendency to separate at the borders of grains. Cu has a favorable enthalpy for mixing with other constitutional elements; however, Cu and Ni have a strong attraction that prevents Cu from completely segregating into the FCC phase, which results in the creation of the F2 phase following SPS.

Figure 5.1(b) shows the XRD patterns for bulk alloys made via the casting process. Si$_0$, Si$_{0.3}$, and Si$_{0.6}$ in the CoCrCuFeNi HEA demonstrate a single FCC phase solid solution. No other phase is generated in the cast alloy samples till 0.9 Si addition, indicating that the FCC phase is a comparatively stable phase in these HEAs at high temperatures. Due to the slower diffusion in the multi-element alloys, the FCC phase survives cooling. In contrast, the intermetallic compound in the CoCrCuFeNiSi0.9 alloy was found and identified as a Ni3Si compound by XRD.

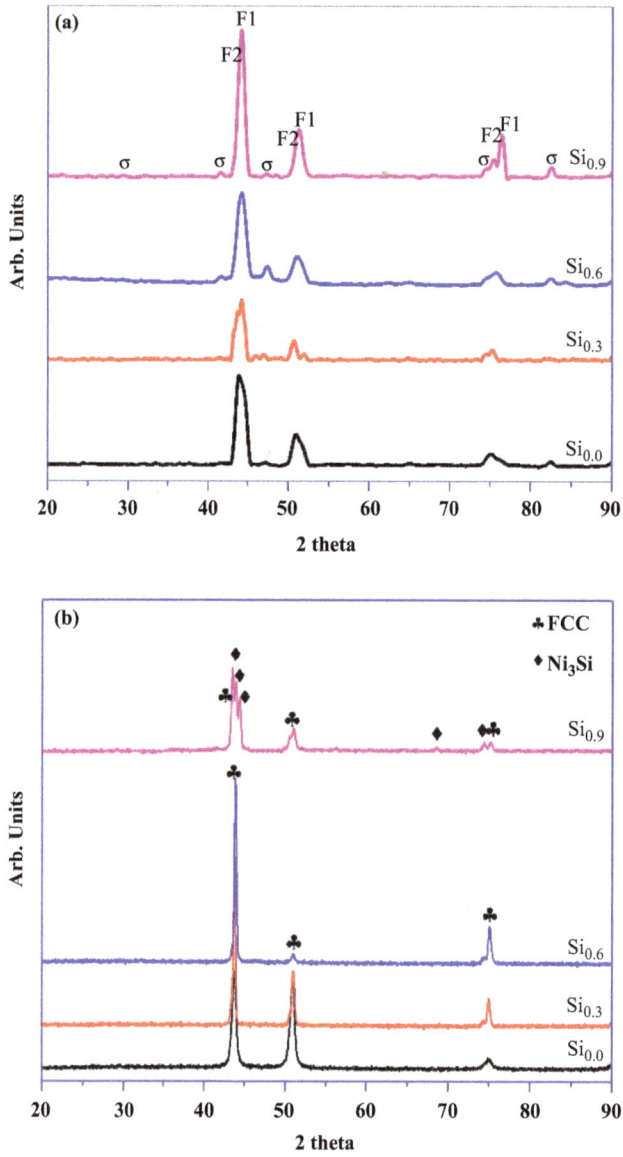

Figure 5.1: The XRD pattern of bulk alloy samples were prepared by: (a) SPS and (b) Arc Melting (AM) [11].

5.4 Microstructure characterization

CoCrCuFeNiSix alloys produced by SPS are depicted in the Back Scattered mode SEM micrographs in Figure 5.2, along with compositional information from the EDS analysis. The creation of two stages is indicated by the gray and black patches in the SEM pictures. While the black contrast zone is Cr-rich, the gray region has a higher Co, Ni, Fe, and Cu content. It can be seen from a comparison of the XRD (Figure 5.1(a)) and the SEM micrographs (Figure 5.2) that the gray section contains two FCC phases. Cu is abundant in the outer layers, whereas Ni, Co, and Fe are abundant in the inner layers. Therefore, it is possible that in the FCC phases, one of the phases, F1, is composed of Fe (Ni, Co), while F2 is composed of Cu (Ni). The **Cr-rich phase** is in the black area. The contrast difference between the F1 and F2 phases is quite small since the atomic numbers of the used elements are relatively close to one another.

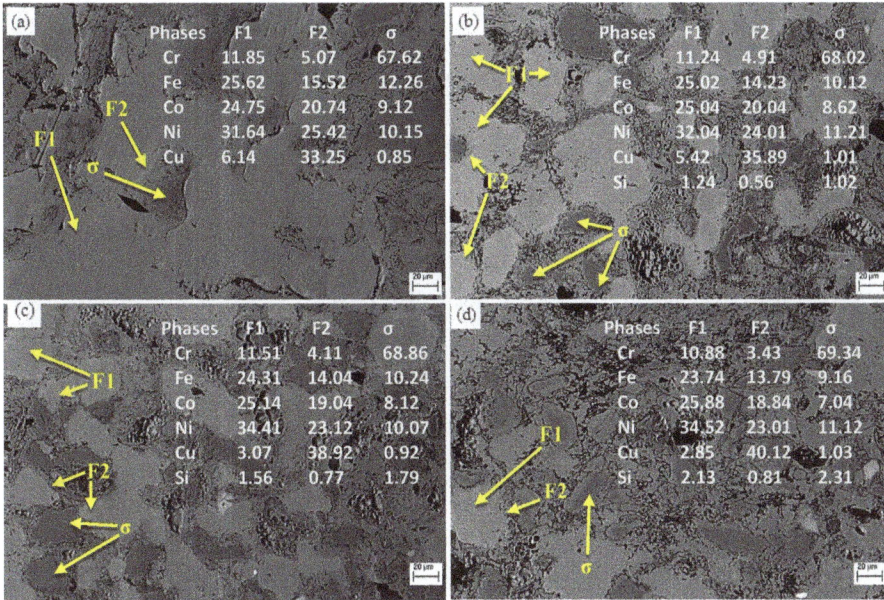

(a)

Phases	F1	F2	σ
Cr	11.85	5.07	67.62
Fe	25.62	15.52	12.26
Co	24.75	20.74	9.12
Ni	31.64	25.42	10.15
Cu	6.14	33.25	0.85

(b)

Phases	F1	F2	σ
Cr	11.24	4.91	68.02
Fe	25.02	14.23	10.12
Co	25.04	20.04	8.62
Ni	32.04	24.01	11.21
Cu	5.42	35.89	1.01
Si	1.24	0.56	1.02

(c)

Phases	F1	F2	σ
Cr	11.51	4.11	68.86
Fe	24.31	14.04	10.24
Co	25.14	19.04	8.12
Ni	34.41	23.12	10.07
Cu	3.07	38.92	0.92
Si	1.56	0.77	1.79

(d)

Phases	F1	F2	σ
Cr	10.88	3.43	69.34
Fe	23.74	13.79	9.16
Co	25.88	18.84	7.04
Ni	34.52	23.01	11.12
Cu	2.85	40.12	1.03
Si	2.13	0.81	2.31

Figure 5.2: The detailed SEM micrographs of $CuCrCoFeNiSi_x$ HEAs prepared by SPS: (a) x = 0.0 (b) x = 0.3(c) x = 0.6 (d) x = 0.9 [11].

The SEM micrographs of the HEAs created during casting are shown in Figure 5.3. All alloys contain a dendritic structure, which is a common casting property. Based on their chemical makeup and the nature of their interactions, the constituents are split between the dendritic structure (DR) and the **inter-dendritic structure** (ID) regions. With a larger Si content, the CoCrCuFeNiSix alloys undergo various and more complex modifications in their microstructure. No discernible difference in microstructures was seen for Si additions of x = 0, 0.3, or 0.6 Figure 5.3(a–c). In contrast, a considerable fluctuation was seen as the Si content increased to 0.9 (Figure 5.3(d). Up to 0.6 of Si, only the FCC phase was seen, according to a careful inspection of the XRD data (Figure 5.1(b) and the SEM micrographs (Figure 5.3); however at Si addition of 0.9, the microstructure contains two phases (FCC and Ni3Si). While Si accumulates in the dendritic area to create the Ni3Si phase, Cu segregation in the interdendritic region evolved as the FCC phase. For the remaining elements (Fe, Ni, Cr, and Co), there hasn't been any discernible segregation in any area. The Cu-rich phase is formed due to the positive mixing enthalpies ($+\Delta H_{mix}$) with the other component elements.

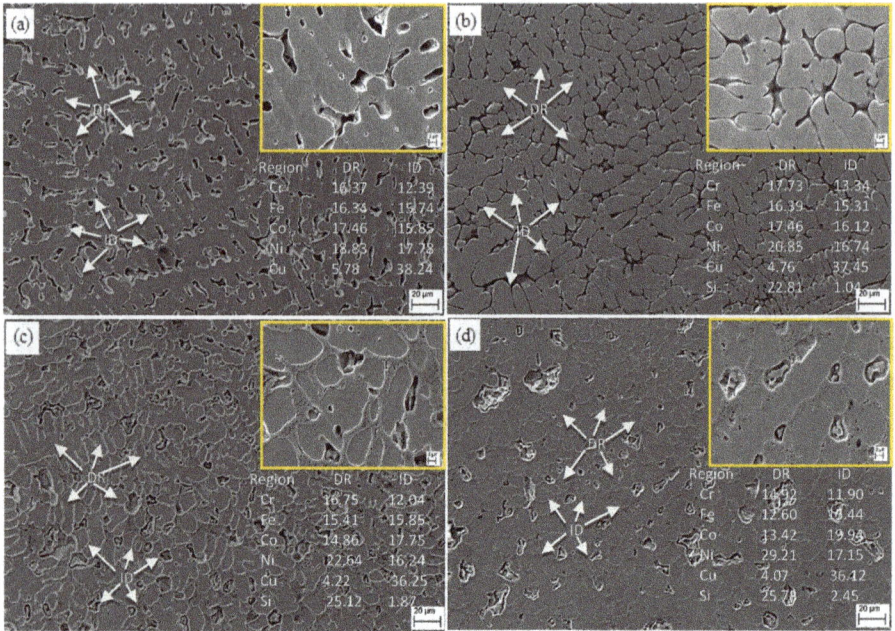

Figure 5.3: The detailed SEM micrographs of CuCrCoFeNiSi$_x$ HEAs prepared by arc melting process: (a) x = 0.0 (b) x = 0.3(c) x = 0.6 (d) x = 0.9 [11].

5.5 Conclusion

A successful series of CoCrCuFeNiSix HEAs was produced using two distinct processing techniques, spark plasma sintering (SPS) and vacuum arc melting (VAM). In samples consolidated by SPS, the XRD data demonstrate the development of two FCC phases, where F1 (NiCoFe) is the first phase and the other phase is F2 (Cu-rich), which combined with other sigma phases (Cr-rich). In contrast, only one FCC phase could be seen in the CoCrCuFeNiSi$_x$ HEAs where the Si addition was up to 0.6 and the samples were made using the vacuum arc melting method. Because Si and Ni components have the largest negative enthalpy of mixing, when Si addition rises from x = 0.6 to 0.9, the Ni3Si phase forms.

References

[1]   J. W. Yeh, et al., "Nanostructured high-entropy alloys with multiple principal elements: Novel alloy design concepts and outcomes," *Advanced Engineering Materials*, vol. 6, no. 5, pp. 299–303, 2004, doi: 10.1002/adem.200300567.

[2]   W. Zhang, P. K. Liaw, and Y. Zhang, "Science and technology in high-entropy alloys," *Science China Materials*, vol. 61, no. 1, pp. 2–22, 2018, doi: 10.1007/s40843-017-9195-8.

[3]   A. Kumar, P. Dhekne, A. K. Swarnakar, and M. K. Chopkar, "Analysis of Si addition on phase formation in AlCoCrCuFeNiSix high entropy alloys," *Materials Letters*, vol. 188, no. October, 2016, pp. 73–76, 2017, doi: 10.1016/j.matlet.2016.10.099.

[4]   S. Vrtnik, et al., "Superconductivity in thermally annealed Ta-Nb-Hf-Zr-Ti high-entropy alloys," *Journal of Alloys and Compounds*, vol. 695, pp. 3530–3540, 2017, doi: 10.1016/j.jallcom.2016.11.417.

[5]   S. Praveen, B. S. Murty, and R. S. Kottada, "Alloying behavior in multi-component AlCoCrCuFe and NiCoCrCuFe high entropy alloys," *Materials Science and Engineering A*, vol. 534, pp. 83–89, 2012, doi: 10.1016/j.msea.2011.11.044.

[6]   C. J. Tong, et al., "Microstructure characterization of AlxCoCrCuFeNi high-entropy alloy system with multiprincipal elements," *Metallurgical and Materials Transactions A: Physical Metallurgy and Materials Science*, vol. 36, no. 4, pp. 881–893, 2005, doi: 10.1007/s11661-005-0283-0.

[7]   W. Ji, et al., "Mechanical alloying synthesis and spark plasma sintering consolidation of CoCrFeNiAl high-entropy alloy," *Journal of Alloys and Compounds*, vol. 589, pp. 61–66, 2014, doi: 10.1016/j.jallcom.2013.11.146.

[8]   Y. P. Wang, B. S. Li, M. X. Ren, C. Yang, and H. Z. Fu, "Microstructure and compressive properties of AlCrFeCoNi high entropy alloy," *Materials Science and Engineering A*, vol. 491, no. 1–2, pp. 154–158, 2008, doi: 10.1016/j.msea.2008.01.064.

[9]   X. W. Qiu, "Microstructure and properties of AlCrFeNiCoCu high entropy alloy prepared by powder metallurgy," *Journal of Alloys and Compounds*, vol. 555, pp. 246–249, 2013, doi: 10.1016/j.jallcom.2012.12.071.

[10] D. G. Shaysultanov, N. D. Stepanov, A. V. Kuznetsov, G. A. Salishchev, and O. N. Senkov, "Phase composition and superplastic behavior of a wrought AlCoCrCuFeNi high-entropy alloy," *Jom*, vol. 65, no. 12, pp. 1815–1828, 2013, doi: 10.1007/s11837-013-0754-5.

[11] A. Kumar, A. K. Swarnakar, A. Basu, and M. Chopkar, "Effects of processing route on phase evolution and mechanical properties of CoCrCuFeNiSix high entropy alloys," *Journal of Alloys and Compounds*, vol. 748, pp. 889–897, Jun. 2018, doi: 10.1016/j.jallcom.2018.03.242.

K. Raja Rao, Satish Pujari, Man Mohan, Manoj S. Choudhary,
Vinay Kumar Soni, Agnivesh Kumar Sinha

Chapter 6
Correlation of the sintering parameters with the mechanical properties of HEAs processed through the powder metallurgy route

Abstract: In order to produce high-entropy alloys (HEAs), researchers have experimented with several processing strategies. HEAs may be produced in a variety of configurations, such as dense solid castings, powder metallurgy components, and films. The techniques for the synthesis of HEA may be grouped into three categories: melting and casting, powder metallurgy, and deposition techniques. These are the broad categories that can be used to describe the processing methods.

Synthesis of HEAs by **solid-state processing** comprises mechanical alloying (MA) of the elemental blends, followed by consolidation. The MA process is a high-energy ball milling of the elemental powder mixes that involves the diffusion of species into each other in order to generate a homogenous alloy. The MA process is also referred to as the ball milling method. One of the benefits of MA is that it may provide great uniformity in the composition of the alloy, which is one of its advantages. In order to produce dense components, the HEAs that have been produced using the powder metallurgy process must be sintered. Conventional sintering of nanocrystalline alloy powders can result in considerable grain growth if the alloy powders are subjected to high temperatures for an extended length of time. In order to counteract this issue, the nanocrystalline alloys that are produced by MA are often sintered in spark plasma before being consolidated (SPS).

Keywords: High-entropy alloys, Powder metallurgy, Mechanical alloying, Conventional sintering, Spark Plasma Sintering

Man Mohan, Department of Mechanical Engineering, Rungta College of
Engineering and Technology, Bhilai, Chhattisgarh 490024, India, e-mail: manmohan.cimt@gmail.com
K. Raja Rao, Satish Pujari, Department of Mechanical Engineering, Lendi Institute of Engineering and
Technology, Vizianagaram, Andhra Pradesh 535005, India
Manoj S. Choudhary, Agnivesh Kumar Sinha, Department of Mechanical Engineering, Rungta College
of Engineering and Technology, Bhilai, Chhattisgarh 490024, India
Vinay Kumar Soni, Department of Mechanical Engineering, Columbia Institute of Engineering and
Technology, Raipur, Chhattisgarh 493111, India

https://doi.org/10.1515/9783110769470-006

6.1 Introduction

In the materials science domain, high-entropy alloys (HEAs) are made up of multi-principal component systems that are classified according to the concentration of their constituent elements, which can range from 5 to 35, at present [1]. Due to their remarkable compositional complexity and microstructural variation, apart from their outstanding mechanical attributes, HEAs have fascinated many researchers since their breakthrough in 2004 by Yeh [2] and Cantor [3].

It is well established that the occurrence and the secondary phase volume fraction in HEAs are substantially influenced by the processing circumstances as well as the post-processing heat treatment used. As a result, the synthesis of bulk HEA has been a prominent area of investigation, with the goal of optimizing its properties. The often used processing procedures are vacuum arc melting and **mechanical alloying** (MA). MA powder is further consolidated and sintered in the final step. Mechanical alloying, as compared to casting, is based on the creation of dislocations, which serve as energy storages along the grain boundaries and provide the additional driving force, which in turn facilitates the formation of a solid solution [4].

Since the last 20 years, the amplification methods have depended on the use of high current intensity, of which the SPS approach is arguably very well-known and has grabbed the curiosity of researchers everywhere. When a substantial quantity of pulsed direct electrical current is applied to metallic or ceramic powder that is restricted in a die (most typically graphite), the process is known as Spark Plasma Sintering (SPS). With its capacity to maintain nanocrystalline microstructures and produce materials with unique functional characteristics, SPS has attracted a great deal of attention in the scientific community. As a result, SPS is a real nonequilibrium processing technology, due to the fact that it involves lower temperatures and reduced processing durations, compared to traditional approaches. SPS has achieved progress in practically all areas of materials with improved characteristics, as a result of decades of study and development. Moreover, SPS is uniquely positioned to benefit from grain boundary strengthening because it has the ability to synthesize materials to a high density while minimizing grain growth.

This was determined by the use of an array of hardness datasets of HEAs for the present work after reviewing a number of previously published publications that were based on experimental work. All the HEAs were synthesized using the **SPS** process, which were then utilized as a dataset to determine hardness. A combination of the all-type hardness range and the sintering temperature has been selected for this process.

6.2 Mechanical properties correlation with sintering parameters of HEA

Due to their use in structural applications, HEAs have to be synthesized in bulk form, with the appropriate microstructure and property combinations. The production of bulk HEAs can be accomplished through a variety of processes, including those that take place in a liquid state (these include arc melting, induction melting, laser melting, and electric resistance melting), and those that take place in a solid state (these include mechanical alloying and sintering). Sintering, on the other hand, offers benefits over the melting approach in that the alloys containing elements that have various melting temperatures and differing densities may be synthesized using the SPS technique without any significant segregation, porosity, or inhomogeneity occurring in the process. SPS involves applying a high current (800–1,500 A) and a high pressure (30–100 MPa) at the same time in order to obtain the necessary dense microstructure with restricted grain development in a controlled environment through the appropriate choice of sintering parameters.

It has been determined that the temperature at which the HEA is sintered as well as the elemental composition of the alloy are crucial parameters that greatly influence the hardness of the material. Furthermore, the sintering parameters have an influence not only on the mechanical characteristics of the sintered pellet but also on the relative density of the alloy. Table 6.1 summarizes some of the results obtained by researchers who developed HEA by the MA and SPS technique. TiZrNbMoTa HEA was produced by M. Laurent Brocq et al. using MA and SPS. During this investigation, they experimented with different sintering temperatures and studied the influence of those temperatures on the mechanical characteristics of the alloys. According to the findings of their research, the hardness value of the material drops as the sintering temperature rises; at the same time, the relative density increases with a rise in temperature. Similar results were obtained in the investigation of Lukasz Rogal et al. [19].

6.3 Conclusions

As a result of this brief study, it can be concluded that a correlation exists between the mechanical properties and the sintering parameters. Furthermore, the following conclusion can be drawn: the hardness of the HEA decreases as the sintering temperature rises, while the relative density of the alloy increases with a rise in the sintering temperature.

Table 6.1: Sintering parameters and mechanical properties of HEAs processed by MA and SPS.

Ref.	HEA system	Sintering temperature (°C)	Sintering pressure (MPa)	Holding time at maximum temp. (min.)	Hardness (HV)	Density (g/cm³) Theoretical (g/cm³)	Actual (g/cm³)	Relative (%)
[5]	TiZrNbMoTa	1,300	50	30	1,162	9.12	8.43	92.43
		1,400	50	30	1,181	9.12	8.68	95.18
		1,500	50	30	1,088	9.12	8.73	95.72
		1,600	50	30	1,040	9.12	8.76	96.05
[6]	CoCrFeMnNi	1,000	45	5	–	–	–	98.12
[7]	CoCrFeMnNi	850	100	4	254.9	–	–	–
[8]	CoCrFeMnNiV	900	60	5	525	–	–	–
[9]	AlCoCrNiTi–C	1,000	60	10	860 ± 7.8	–	–	–
[10]	AlCuNiFeCr	700	150	15	490	5.4% (Porosity)		
		800	150	5	724	1.1% (Porosity)		
		800	150	15	846	0.6% (Porosity)		
		900	150	15	867	0.6% (Porosity)		
[11]	TiAlMoSiW	800	50	8	630	–	–	94.0
		900	50	8	680	–	–	95.9
		1,000	50	8	803	–	–	96.6

Ref	Composition							
[12]	$Al_{35}Cr_{14}Mg_{96}Ti_{35}V_{10}$	750	50	5	460	4.05	–	85
[13]	AlCoCrFeNi	1,000	80	–	525	–	–	91.1
[14]	MgAlSiCrFeNi	800	50	15	1,018 (9.98 GPa)	5.09	5.06	99.40
[15]	$FeCoCrNiMoSi_{0.5}$	1,150	–	–	725	–	–	–
	$FeCoCrNiMoSi_{1.0}$	1,150	–	–	790	–	–	–
	$FeCoCrNiMoSi_{1.5}$	1,150	–	–	1,186	–	–	–
[16]	NiCoCrFe	900	45	9	622	–	–	–
	$NiCoCrFeZr_{0.4}$	900	45	9	828	–	–	–
[17]	NbTaTiZr	700	30	–	607.8	–	8.79	–
	NbTaTiZr	800	30	–	662	–	8.84	–
	NbTaTiZr	900	30	–	559	–	8.87	–
	NbTaTiZr	1,000	30	–	543	–	8.88	–
	NbTaTiZr	1,100	30	–	529	–	8.87	–
[18]	FeCoCrNiMo	950	35	10	474	8.286	8.09	96.14
[19]	$Al_{25}Co_{25}Cr_{25}Fe_{25}$	1,050	35	15	990	–	–	–
	$Al_{20}Co_{20}Cr_{20}Fe_{20}Ni_{20}$	1,050	35	15	710	–	–	–
	$Al_{10}Co_{30}Cr_{20}Fe_{35}Ni_5$	1,050	35	15	516	–	–	–
	$Al_{15}Co_{30}Cr_{15}Fe_{40}Ni_5$	1,050	35	15	697	–	–	–

(continued)

Table 6.1 (continued)

Ref.	HEA system	Sintering temperature (°C)	Sintering pressure (MPa)	Holding time at maximum temp. (min.)	Hardness (HV)	Density (g/cm³) Theoretical (g/cm³)	Actual (g/cm³)	Relative (%)
[20]	$Fe_{24.1}Co_{24.1}Cr_{24.1}Ni_{24.1}Mo_{3.6}$	1,150	30	8	440	–	–	–
		1,150	35	8	320	–	–	–
		1,150	40	8	245	–	–	–
[21]	AlCrCuFeZn	800	50	5	627		5.9	91.3

References

[1]   K. R. Rao and S. K. Sinha, "Effect of sintering temperature on microstructural and mechanical properties of SPS processed CoCrCuFeNi based ODS high entropy alloy," *Materials Chemistry and Physics*, vol. 256, no. August, p. 123709, 2020, doi: 10.1016/j.matchemphys.2020.123709.

[2]   J. W. Yeh, et al., "Nanostructured high-entropy alloys with multiple principal elements: Novel alloy design concepts and outcomes," *Advanced Engineering Materials*, vol. 6, no. 5, pp. 299–303, 2004, doi: 10.1002/adem.200300567.

[3]   B. Cantor, I. T. H. Chang, P. Knight, and A. J. B. Vincent, "Microstructural development in equiatomic multicomponent alloys," *Materials Science and Engineering A*, vol. 375–377, no. 1–2, SPEC. ISS., pp. 213–218, 2004, doi: 10.1016/j.msea.2003.10.257.

[4]   K. Raja Rao and S. K. Sinha, "Phase evolution in novel Y2O3 dispersed CrCuFeNiZn nanocrystalline multicomponent alloy prepared by mechanical alloying," *Vacuum*, vol. 184, no. August, p. 109802, 2021, doi: 10.1016/j.vacuum.2020.109802.

[5]   C. Zhu, Z. Li, C. Hong, P. Dai, and J. Chen, "Microstructure and mechanical properties of the TiZrNbMoTa refractory high-entropy alloy produced by mechanical alloying and spark plasma sintering," *International Journal of Refractory Metals and Hard Materials*, vol. 93, no. March, p. 105357, 2020, doi: 10.1016/j.ijrmhm.2020.105357.

[6]   F. Jiang, et al., "In-situ formed heterogeneous grain structure in spark-plasma-sintered CoCrFeMnNi high-entropy alloy overcomes the strength-ductility trade-off," *Materials Science and Engineering A*, vol. 771, no. October, 2019, p. 138625, 2020, doi: 10.1016/j.msea.2019.138625.

[7]   M. Laurent-Brocq, et al., "Microstructure and mechanical properties of a CoCrFeMnNi high entropy alloy processed by milling and spark plasma sintering," *Journal of Alloys and Compounds*, vol. 780, pp. 856–865, 2019, doi: 10.1016/j.jallcom.2018.11.181.

[8]   M. Vaidya, A. Karati, K. Guruvidyathri, M. Nagini, K. G. Pradeep, and B. S. Murty, "Suppression of σ-phase in nanocrystalline CoCrFeMnNiV high entropy alloy by unsolicited contamination during mechanical alloying and spark plasma sintering," *Materials Chemistry and Physics*, vol. 255, no. May, p. 123558, 2020, doi: 10.1016/j.matchemphys.2020.123558.

[9]   R. Anand Sekhar, A. S. Shifin, and N. Firoz, "Microstructure and mechanical properties of AlCoCrNiTi–C High Entropy Alloy processed through Spark Plasma Sintering," *Materials Chemistry and Physics*, vol. 270, no. April, p. 124846, 2021, doi: 10.1016/j.matchemphys.2021.124846.

[10]   A. I. Yurkova, V. V. Cherniavsky, V. Bolbut, M. Krüger, and I. Bogomol, "Structure formation and mechanical properties of the high-entropy AlCuNiFeCr alloy prepared by mechanical alloying and spark plasma sintering," *Journal of Alloys and Compounds*, vol. 786, pp. 139–148, 2019, doi: 10.1016/j.jallcom.2019.01.341.

[11]   L. Rudolf Kanyane, A. P. I. Popoola, N. Malatji, and M. B. Shongwe, "Spark plasma sintering consolidation of equi-atomic tialmosiw high entropy alloy," *Procedia Manufacturing*, vol. 35, pp. 968–973, 2019, doi: 10.1016/j.promfg.2019.06.043.

[12]   P. Chauhan, S. Yebaji, V. N. Nadakuduru, and T. Shanmugasundaram, "Development of a novel light weight Al35Cr14Mg6Ti35V10 high entropy alloy using mechanical alloying and spark plasma sintering," *Journal of Alloys and Compounds*, vol. 820, pp. 153367, 2020, doi: 10.1016/j.jallcom.2019.153367.

[13]   A. Fourmont, S. Le Gallet, O. Politano, C. Desgranges, and F. Baras, "Effects of planetary ball milling on AlCoCrFeNi high entropy alloys prepared by Spark Plasma Sintering: Experiments and molecular dynamics study," *Journal of Alloys and Compounds*, vol. 820, pp. 153448, 2020, doi: 10.1016/j.jallcom.2019.153448.

[14]   N. Singh, Y. Shadangi, V. Shivam, and N. K. Mukhopadhyay, "MgAlSiCrFeNi low-density high entropy alloy processed by mechanical alloying and spark plasma sintering: Effect on phase evolution and

thermal stability," *Journal of Alloys and Compounds*, vol. 875, pp. 159923, 2021, doi: 10.1016/j. jallcom.2021.159923.

[15] Y. Yang, et al., "Microstructure and properties of FeCoCrNiMoSix high-entropy alloys fabricated by spark plasma sintering," *Journal of Alloys and Compounds*, vol. 884, p. 161070, 2021, doi: 10.1016/j. jallcom.2021.161070.

[16] P. Moazzen, M. R. Toroghinejad, T. Zargar, and P. Cavaliere, "Investigation of hardness, wear and magnetic properties of NiCoCrFeZrx HEA prepared through mechanical alloying and spark plasma sintering," *Journal of Alloys and Compounds*, vol. 892, p. 161924, 2022, doi: 10.1016/j. jallcom.2021.161924.

[17] T. Xiang, Z. Cai, P. Du, K. Li, Z. Zhang, and G. Xie, "Dual phase equal-atomic NbTaTiZr high-entropy alloy with ultra-fine grain and excellent mechanical properties fabricated by spark plasma sintering," *Journal of Materials Science and Technology*, vol. 90, pp. 150–158, 2021, doi: 10.1016/j. jmst.2021.03.024.

[18] Y. B. Peng, et al., "Microstructures and mechanical properties of FeCoCrNi-Mo High entropy alloys prepared by spark plasma sintering and vacuum hot-pressed sintering," *Materials Today Communications*, vol. 24, no. November, 2019, 2020, doi: 10.1016/j.mtcomm.2020.101009.

[19] Ł. Rogal, et al., "Microstructure and mechanical properties of Al–Co–Cr–Fe–Ni base high entropy alloys obtained using powder metallurgy," *Metals and Materials International*, vol. 25, no. 4, pp. 930–945, 2019, doi: 10.1007/s12540-018-00236-5.

[20] M. Zhang, et al., "Gradient distribution of microstructures and mechanical properties in a fecocrnimo high-entropy alloy during spark plasma sintering," *Metals (Basel)*, vol. 9, no. 3, 2019, doi: 10.3390/met9030351.

[21] K. R. Cardoso, B. D. S. Izaias, L. D. S. Vieira, and A. M. Bepe, "Mechanical alloying and spark plasma sintering of AlCrCuFeZn high entropy alloy," *Materials Science and Technology (United Kingdom)*, vol. 36, no. 17, pp. 1861–1869, 2020, doi: 10.1080/02670836.2020.1839195.

Rituraj Chandrakar, K. Sridhar, Prem Shankar Sahu,
Saurabh Chandraker, Pankaj Kumar Gupta

Chapter 7
Basic alloying elements used in high-entropy alloys

Abstract: The mechanical characteristics of high-entropy alloys (HEAs) can be improved by a variety of alloying elements; however, it is unclear how the alloying of various elements affects the changes in the microstructure and the mechanical properties
of HEAs. The alloying elements like Cr, V, Ti, Zr, and Hf regulate the melting temperature, lattice constant, and the mass density of HEAs. The electrical structure and the
mechanical characteristics of HEAs are significantly impacted by the valence electron
concentration. High VEC can enhance mechanical characteristics while decreasing its
ductility. Ti significantly affects ductility, while Cr-alloying significantly affects the mechanical characteristics of HEAs. Our findings offer the fundamental understanding required to direct the development of HEAs with superior mechanical characteristics.

Keywords: Lattice constant, Valence electron concentration, Crystal structure, Alloy,
Entropy

7.1 Introduction

Alloys typically include only one main element and small additions of additional elements to change their properties. HEAs are alloys made up of five or more metals in
equal or almost equal amounts.

Due to their potentially useful features, HEAs are presently the center of intense
interest in metallurgical, mechanical, and material science fields. An HEA is created
by melting five or more metals in almost equal amounts to create stable random solid

Rituraj Chandrakar, Department of Mechanical Engineering, NMDC DAV Polytechnic College Geedam,
Dantewada, Chhattisgarh 494441, India, e-mail: rituraj.chandraker@gmail.com; And Department of Mechanical
Engineering, National Institute of Technology Karnataka (NITK), Surathkal, Mangalore, Karnataka 575025, India
K. Sridhar, Department of Mechanical Engineering, Lendi Institute of Engineering and Technology, Viziana-
garam, Andhra Pradesh 535005, India
Prem Shankar Sahu, Department of Mechanical Engineering, Bhilai Institute of Technology, Durg, Chhattis-
garh 491001, India
Saurabh Chandraker, Department of Mechanical Engineering, National Institute of Technology Karnataka
(NITK), Surathkal, Mangalore, Karnataka 575025, India
Pankaj Kumar Gupta, Department of Mechanical Engineering, Guru Ghasidas Vishwavidyalaya, Koni,
Bilaspur, Chhattisgarh 495009, India

https://doi.org/10.1515/9783110769470-007

solutions. High mixing entropy promotes the creation of phases that are similar to solutions and, generally, results in a microstructure that is simpler.

The name comes from the discovery that these alloys' extraordinarily high mixing entropy (ΔH_{mix}) favors stability, or the ability to maintain their microstructures without extrication into various phases through ordering or segregation, as is the case with conventional alloys. Electrical, magnetic, and high-temperature applications call for parts with good oxidation and corrosion resistance at high temperatures. Therefore, the utilization of many components in the construction of these novel materials marks their significant distinction from conventional alloys. Currently, single-phase, disordered solid solution alloys are favored in HEA investigations.

The last ten years, however, have seen a significant amount of research on single-phase HEAs. These are cubic forms, with a single phase that are **face- and body-centered**. Here, the BCC structure delivers great strength whereas the FCC structure offers improved ductility. On the other hand, a novel composite structural (soft and ductile FCC and intermetallic phase usually hard) HEA must be developed in order to achieve high ductility along with high strength. Very little study had been conducted recently on HEAs to create composite structures. According to a previous study of the literature, the development of "Al-Ni–based" intermetallic alloy with a soft and ductile FCC phase was the main goal [1, 2].

It is thought that solid solution strengthening in HEAs is more effective than in traditional alloys. Numerous HEA microstructures, including single phase, multiphase, intermetallic, nano-crystalline, and even non **crystalline** (amorphous) alloys, have been created [3–7].

7.2 Selection of main constituent elements in HEAs

Table 7.1 lists the elements that can be utilized to create HEAs. The primary components of HEAs are non-metallic elements, including C, B, Si, P, and S, as well as metallic elements that can be chosen from the metallic group.

Table. 7.1: Major constituent elements for possible use in HEAs.

Major metallic elements	Li, Cr, Be, Mg, V, Ti, Sc, Al, Fe, Zn, Cu, Ni, Co, Zr, Nb, Mo, Sm, Au, Eu, Gd, Tb, Rh, Pb, Pd, Ag, Nd, W, Ta, hf, pt
Minor metallic elements	Ga, Ge, Sy, Cd, In, Sn, Sb, Ru, Bi, La, Ce, Pr
Major non-metallic elements	C, B, Si, P, S, O, N

Jien-Wei et al. demonstrated this by using the CuCoNi-CrAlFe alloy system as an example to show how an arbitrary selection of a set of 13 metallic elements that are mutually miscible allows the design of 7,099 HEA systems, with 5 to 13 constituent elements in equiatomic or in equimolar ratios. Thus, even with components that cannot induce chemical immiscibility, many alloys can be created. Cantor showed a variety of researched alloys, including all unitary, the majority binary, a few ternary, and additional alloy systems. These elements are chosen for alloying:

–   From the literature and standards, list all the metals used for structural purposes. Remove all non-metals, starting with the elements of the periodic table.
–   List desired characteristics
–   To locate atoms with an atomic size difference of no more than 8%, apply the Hume-Rothery criteria. It is crucial for cast alloys.

The constituent major alloying elements must have similar properties, including matching atomic sizes and electronegativity. However, it appears that the production of solid solutions in HEAs does not comply with the Hume-Rothery rule. For instance, it cannot explain "how the addition of an FCC element such as Al can finally convert the FCC-type CoCrCuFeNi to a BCC structure or why the equi-atomic Co(HCP)-Cr(BCC)-Cu(FCC)-Fe (bcc)-Ni(FCC) alloy produces an FCC-typed solid solution" [1, 2]. These phase compositions in multiphase HEAs are distant from the solute portions where the substitutional solid solutions may form in accordance with the respective binary and ternary phase diagrams, if the effects of the additional minor element can be ignored. Consequently, the cause of the thermodynamic instability is the precipitation caused by ion exposure, which is comparable to the behaviour of precipitation in many immiscible binary systems where the associated binary and ternary compositional phase diagrams can be used to generate solid solutions.

7.3 Benefits of high-entropy alloys

The stimulation of research into previously unconsidered compositionally complicated alloys is one of the main advantages of HEAs.

Another significant advantage of HEAs is that they offer a technique to create a huge variety of new alloys. There are a large variety of compositions that may be created simply by experimenting with different conceivable combinations, each of which has at least five metal elements. The majority of them have a high likelihood of producing usable alloys, offering significant potential for key scientific and practical discoveries. It has been demonstrated that a huge number of systems are possible, from which only the most valuable will be explored, even if one restricts the amount of new alloy system to be studied to those created from totally miscible metals [8].

7.4 Necessity for novel alloys in the industry

There is still a critical need for material advancement in a variety of applications, including [9]:

1. Better **hot corrosion** (sulfidation) resistance, oxidation resistance, and elevated-temperature strength, which are all attributes of engine materials.
2. Nuclear materials have lower neutron absorption and better strength at high temperatures.
3. Better high-temperature strength, high-impact strength, better wear resistance, high corrosion resistance, high oxidation resistance, low friction, and nonstick properties are found in tool materials.
4. Waste incinerator: improved wear, corrosion, oxidation resistance, and elevated-temperature strength.
5. Refractory construction frames have superior strength at elevated temperatures to withstand firing.
6. Better mechanical properties such as strength, hardness, creep resistance, toughness, and workability in light transportation materials.
7. High electrical property, such as resistance and high magnetic permeability (usually above 3 GHz) are characteristics of high-frequency communication materials.
8. Decorative coatings: improved wear resistance, nonstick, anti-fingerprint, and anti-bacterial coatings.
9. Low-cost, highly reversible volumetric and gravimetric hydrogen storage materials for mobile devices with near-ambient cycling conditions.
10. Superconductor: greater critical current and temperature. Higher thermoelectric figure of merit for thermoelectric materials.
11. Higher strength and resilience in the golf club head.

7.5 Future scope for high-entropy alloys

The transportation and energy sectors use HEAs because of their low density and great strength. Performance, dependability, and endurance under harsh working circumstances are requirements for these applications [10]. HEAs make excellent replacements for steel- and titanium-based alloys. Other uses are such as aero-engine compressor blades, which are commonly made from Ti-based alloys. HEAs can be fabricated into rod-shaped and powder particles and then plasma arcing or thermally spraying them onto the surface of machine tools and other mechanical components, for which hardfacing technology can be used [11]. In the **hardfacing** process, thermal spray welding is used to apply a thick layer of a material that is resistant to wear and/or corrosion. Molds, dies, tools, and nozzles are frequently used in industrial settings [12]. They can be employed, particularly in electronics, to reduce electromagnetic interference. For instance, a coating of 1 m can function as a screen in commercial applications at 13,000 MHz. Due to their ability to resist corrosion, oxidation, and wear, HEAs can be utilized as coatings for cookware and food preservation. It is possible to stop bacteria from procreating and forming colonies, such as E. coli. As a result, there are more applications for HEAs.

7.6 Conclusions

The development of HEAs throws up new opportunities in the study of materials science and metallurgy. When compared to traditional alloys, these alloys exhibit exceptional capabilities and distinct traits. Many expensive alloys that are now used in the manufacturing industry can be replaced by HEAs. In HEAs, there is a lot of room for additional study that could result in the creation of novel alloys.

References

[1] S. Wu, Y. Pan, J. Lu, N. Wang, W. Dai, and T. Lu, "Effect of the addition of Mg, Ti, Ni on the decoloration performance of AlCrFeMn high entropy alloy," *Journal of Materials Science and Technology*, vol. 35, 2019, pp. 1629–1635, doi: https://doi.org/10.1016/j.jmst.2019.03.025.

[2] A. Kumar, A. K. Swarnakar, and M. Chopkar, "Phase evolution and mechanical properties of AlCoCrFeNiSix high-entropy alloys synthesized by mechanical alloying and spark plasma sintering," *Journal of Materials Engineering and Performance*, vol. 27, 2018, pp. 3304–3314, doi: https://doi.org/10.1007/s11665-018-3409-4.

[3] A. Kumar and M. Chopkar, "Effects of mutual interaction between constituent elements on phase formation of high entropy alloys," *Journal of Materials Science and Nanotechnology*, vol. 5, 2017, pp. 2–7, doi: https://doi.org/10.15744/2348-9812.5.201.

[4] Y. Zhang, T. T. Zuo, Z. Tang, M. C. Gao, K. A. Dahmen, P. K. Liaw, and Z. P. Lu, "Microstructures and properties of high-entropy alloys," *Progress in Materials Science*, vol. 61, 2014, pp. 1–93, doi: https://doi.org/10.1016/j.pmatsci.2013.10.001.

[5] N. N. Guo, L. Wang, L. S. Luo, X. Z. Li, R. R. Chen, Y. Q. Su, J. J. Guo, and H. Z. Fu, "Effect of composing element on microstructure and mechanical properties in Mo-Nb-Hf-Zr-Ti multi-principle component alloys," *Intermetallics*, vol. 69, 2016, pp. 13–20, doi: https://doi.org/10.1016/j.intermet.2015.10.011.

[6] A. Kumar, P. Dhekne, A. K. Swarnakar, and M. K. Chopkar, "Analysis of Si addition on phase formation in AlCoCrCuFeNiSix high entropy alloys," *Materials Letters*, vol. 188, 2017, pp. 73–76, doi: https://doi.org/10.1016/j.matlet.2016.10.099.

[7] S. Praveen, B. S. Murty, and R. S. Kottada, "Alloying behavior in multi-component AlCoCrCuFe and NiCoCrCuFe high entropy alloys," *Materials Science and Engineering A*, vol. 534, 2012, pp. 83–89, doi: https://doi.org/10.1016/j.msea.2011.11.044.

[8] D. B. Miracle, J. D. Miller, O. N. Senkov, C. Woodward, M. D. Uchic, and J. Tiley, "Exploration and development of high entropy alloys for structural applications," *Entropy*, vol. 16, 2014, pp. 494–525, doi: https://doi.org/10.3390/e16010494.

[9] L. S. Zhang, G. L. Ma, L. C. Fu, and J. Y. Tian, "Recent progress in high-entropy alloys," *Advanced Materials Research*, vol. 631–632, 2013, pp. 227–232, doi: https://doi.org/10.4028/www.scientific.net/AMR.631-632.227.

[10] J. W. Yeh, S. K. Chen, S. J. Lin, J. Y. Gan, T. S. Chin, T. T. Shun, C. H. Tsau, and S. Y. Chang, "Nanostructured high-entropy alloys with multiple principal elements: Novel alloy design concepts and outcomes," *Advanced Materials Research*, vol. 6, 2004, pp. 299–303, doi: https://doi.org/10.1002/adem.200300567.

[11] M. E. Glicksman, Nucleation Catalysis, 2012. doi: https://doi.org/10.1007/978-1-4419-7344-3_12.

[12] Y. Zhang, T. T. Zuo, Z. Tang, M. C. Gao, K. A. Dahmen, P. K. Liaw, and Z. P. Lu, "Microstructures and properties of high-entropy alloys," *Progress in Materials Science*, vol. 61, 2014, pp. 1–93, doi: https://doi.org/10.1016/j.pmatsci.2013.10.001.

Agnivesh Kumar Sinha, Ram Krishna Rathore, Vinay Kumar Soni,
K. Raja Rao, Nitin Upadhyay, Gulab Pamnani

Chapter 8
Effect of alloying elements on the phases of high-entropy alloys

Abstract: High-entropy alloys (HEAs) are acclaimed for their remarkable properties like high strength, high temperature strength, corrosion resistance, irradiations, etc. These properties are realized due to the synergic effect of constituent alloying elements of HEAs, and the end-effect of alloying element is obtained as the phases or microstructure formation of HEAs. The properties of HEAs are the function of phases. Thus, this chapter deals with the influence of alloying elements and their variations on the phases or microstructure of HEAs. It also elucidates the research gaps in the past literature, which would enable the researchers to design tailored HEAs with desired properties.

Keywords: High-entropy alloys, Phases, Microstructure, Properties, Alloying elements

8.1 Introduction

High-entropy alloys (HEAs) are alloys with high entropy of mixing, which results in the formation of the solid solution [1]. Solid solution strengthens the alloys formed [2], and hence, HEAs exhibit remarkable properties namely, strength, corrosion resistance, oxidation resistance, wear resistance, and hardness [3]. Properties of alloy majorly depend on the alloying elements and their compositions, and these elements result in formation of various phases or structures, depending on the respective contents of alloying elements. Elements like Fe, Cu, Cr, Ni, Co, Mo, V, Sn, Ti, Ta, W, Si, Mn, and Nb are often used as alloying elements in HEAs in various compositions.

Agnivesh Kumar Sinha, Ram Krishna Rathore, Mechanical Engineering Department, Rungta College of Engineering and Technology, Bhilai, Chhattisgarh 490024, India, e-mail: sinhaagnivesh@yahoo.in
Vinay Kumar Soni, Mechanical Engineering Department, Columbia Institute Engineering and Technology, Raipur, Chhattisgarh 493111, India
K. Raja Rao, Department of Mechanical Engineering, Lendi Institute of Engineering and Technology, Vizianagaram, Andhra Pradesh 535005, India
Nitin Upadhyay, Department of Mechanical Engineering, Madhav Institute of Technology& Science, Gwalior, Madhya Pradesh 474005, India
Gulab Pamnani, Department of Mechanical Engineering, Malaviya National Institute of Technology, Jaipur, Rajasthan 302017, India

https://doi.org/10.1515/9783110769470-008

A research based on the CoCrFeMnNi HEA [4] showed that the incorporation of Si and Ti showed increment in the strength of HEA. However, it was noticeable that the addition of Ti in CoCrFeMnNi HEA resulted in higher increase in strength than that of CoCrFeMnNiSi HEA. It was also observed from the results that the incorporation of Si stabilized A13 (βMn) phase, whereas addition of Ti stabilized laves phase, σ phase, and A12 (αMn) phase. The solubility of the alloying element in HEA also affects the structure or phase formation, which tends to influence its mechanical properties. In research [5], it was manifested that the addition of Cr in FeCoNiHEA resulted in precipitation (hard phase) at the grain boundaries, which might result in the strengthening of HEA, thus affecting its mechanical properties. It is also reported in a study [6] that $CoCrFeNiW_x$ at $x = 0$ exhibited only FCC structure, and on increasing the W content to $x = 0.2$, $CoCrFeNiW_{0.2}$ exhibited FCC + μ structure. Further addition of W in $CoCrFeNiW_x$ (at $x = 0.7$ and 1.0) showed FCC + μ + BCC structure. Formation of dual-phase structure contributed in enhancing the strength of HEA, while μ phase promoted brittleness.

Literature shows that alloying elements and their composition significantly contribute in deciding the phase, and hence, the mechanical attributes of HEAs. Therefore, this chapter deals with the influence of alloying elements and their composition on the phases or structures of HEAs. This will facilitate in predicting the phases of various HEAs, prior to experimentation, which would conserve resources.

8.2 Effect of alloying elements on phases of HEAs

It would be quite beneficial to know the phases of HEA before its development, as this would help in designing and developing HEAs with desired properties, without any trials or preliminary experiments. This would lead to reduction in time for development of new HEA systems or compositions. In other words, it would be possible to maximize the exploration of the infinite possibilities of compositions for desired applications. This will enable the researchers to come up with the novel, high-performing HEA systems or compositions at a higher rate, and hence, would provide solutions to many unanswered problems.

8.2.1 Face-centered cubic (FCC) phase

HEAs exhibiting FCC phase are known to have some inherent properties that would prove beneficial for several applications. FCC phase exhibited by HEAs usually has good ductility and toughness and lower magnitude of hardness. In general, the alloying elements like Ni and Cu are the promoters of FCC phase in HEAs. It was reported that [7] incorporation of Co in the range of 0.5 to 1.5 atomic wt.% as **alloying element** in $AlCo_xCrFeMo_{0.5}Ni$ HEA reduces the σ phase, thereby increasing BCC phase. However, on further addition of Co beyond 2.0 atomic wt.%, HEA manifests FCC phase, thus proving that addition of Co alloying element in HEAs stabilizes FCC phase. FeCrNiMn HEA-exhibited FCC phase was transformed into FCC + σ phase by addition of Mo as alloying element, which resulted in enhancing the properties of HEAs [8].

In research [9], $(CoFe_2NiV_{0.5}Mo_{0.2})_{100-x}Nb_x$ $(0 \leq x \leq 12)$ manifested FCC phase along with laves phase at $x = 9.0$. Results showed that increase in Nb content in HEAs increases the formation of laves phase, which tends to reduce ductility as trade-off for the strength. Studies also reported that addition of Cu as alloying element in $AlCrFeNiTiCu_x$ promoted the formation of FCC phase, which was BCC, prior to its addition [10]. Also, increase in Cu content resulted in segregation of other alloying elements. Cr and Fe were segregated at **interdendrite** and dendrite regions, whereas Ti, Ni, and Al were segregated only in dendrite region. Mo as alloying element in $FeCoNiCuMo_x (x = 0.2, 0.4, 0.6, 0.8, 1.0)$ HEAs [11] contributed in the formation of FCC phase when incorporated in the range of $x = 0.2$ to $x = 0.6$ atomic wt.%. On further increasing the Mo content, Cu segregation phenomenon was observed, which resulted in formation of FCC1 + FCC2 phase. It was noticeable that hardness, strength, and ductility, all increased with the addition of Mo in $FeCoNiCuMo_x$ HEAs, which was remarkable.

8.2.2 Body-centered cubic (BCC) phase

HEAs with BCC phase are known to exhibit excellent hardness and are usually brittle in nature, which, in turn, reduces their ductility, and hence, **toughness** too. Al, Cr, W, and Si as alloying elements are often seen to manifest the formation of BCC phase in HEAs. It is observed from the literature based on investigation of the phase formation of $Al_xCoCrFeMo_{0.5}Ni$ HEA, [12] wherein Al was varied from $x = 0$ to $x = 2$ in the steps of 0.5 atomic wt.%. It was observed that with the increase in Al content, FCC + σ phase of $Al_0CoCrFeMo_{0.5}Ni$ HEA transformed into BCC1 + BCC2 for $Al_2CoCrFeMo_{0.5}Ni$ HEA (i.e., at $x = 2.0$), and due to this transformation of phase, $Al_2CoCrFeMo_{0.5}Ni$ HEA exhibited almost three times the hardness of $Al_0CoCrFeMo_{0.5}Ni$ HEA, which is remarkable. This study manifested the capability of Al to promote BCC formation in HEAs.

Addition of Ti [13] as alloying element in $CoCrMoNbTi_x$ improved strength of HEAs at the expense of ductility. At $x = 0.2$, HEA exhibited BCC phase only. However, at $x = 0, 0.2, 0.4, 0.5, 1.0$, BCC phase was accompanied by laves phases too. The formation of hard laves phase enhances the strength and the hardness of HEAs. Interestingly, a research study reported that the increase in Mo content as alloying element in $Co_{30}Cr_{10}Fe_{10}Al_{18}Ni_{32-x}Mo_x$ enhanced the volume fraction of BCC phase, wherein HEAs exhibited FCC + BCC phase. Here, Mo also contributed in solid-solution strengthening of HEA, which improved its tensile strength up to 1,250 MPa. Table 8.1 tabulates the phases exhibited by various HEAs along with their corresponding properties, namely, compressive strength, compressive strain, and hardness.

Table 8.1: Phase or structure of HEAs along with their properties.

HEAs	Phase/ structure	Compressive strength (MPa)	Compressive strain (%)	Hardness (HV)	References
CoCrCuFeNi + 1 wt. Y_2O_3	fcc1 + fcc2	1,555	–	468 ± 5	[14]
AlCoCrFeNi	fcc + bcc + σ	1613 ± 22	–	581 ± 6	[15]
AlCoCrCuFeNi	fcc + bcc	–	–	650	[16]
CrCuFeNiZn + 1 wt. Y_2O_3	fcc + bcc	–	–	–	[17]
AlCrFeMnNi + 1 wt. % PSZ	fcc + bcc	–	–	–	[18]
CoCrCuFeNiMo	fcc + bcc	1,624	–	475.09	[19]
CoCrCuFeNiMoAl$_{0.32}$	fcc + bcc	1,486	–	521.75	
CoCrCuFeNiMoAl$_{0.67}$	bcc + Ni3Al	1646.5	–	535.37	
CoCrCuFeNiMoAl	ordered bcc	1,826	–	619.73	
CoCrNiCuMo	fcc + bcc + laves	1,463	–	540.5	[20]
CoCrNiCuMoAl$_{0.26}$	fcc + bcc + laves	1,735	–	572.6	
CoCrNiCuMoAl$_{0.56}$	fcc + bcc + laves	1,750	–	613.4	
CoCrNiCuMoAl	fcc + bcc + laves	1,775	–	621.5	
AlCrFeNiTi	bcc1 + bcc2	–	–	434	[10]
AlCrFeNiTiCu$_{0.5}$	fcc + bcc1 + bcc2	–	–	382	
AlCrFeNiTiCu	fcc + bcc1 + bcc2	–	–	358	
AlCrFeNiTiCu$_{1.5}$	fcc + bcc1 + bcc2	–	–	320	
AlCrFeNiTiCu$_2$	fcc + bcc1 + bcc2	–	–	280	
Al$_{0.4}$Co$_{0.5}$FeNi	fcc + bcc	2,000	50	–	[21]
Al$_{0.4}$Co$_{0.5}$V$_{0.2}$FeNi	fcc + bcc	3,000	60	–	
Al$_{0.4}$Co$_{0.5}$V$_{0.4}$FeNi	fcc + bcc	3,000	60	–	
Al$_{0.4}$Co$_{0.5}$V$_{0.6}$FeNi	fcc + bcc	1,746	36.4	–	

Table 8.1 (continued)

HEAs	Phase/ structure	Compressive strength (MPa)	Compressive strain (%)	Hardness (HV)	References
CoCrFeNiB	fcc + M_3B	895	4.9	–	[22]
CoCrFeNiY$_{0.1}$B	fcc + M_3B	1,350	15	–	
CoMnFeNiAl$_{0.3}$Cu$_{0.7}$B	fcc + M_3B	1,205	19.4	–	
CoMnFeNiAl$_{0.3}$Cu$_{0.7}$Y$_{0.1}$B	fcc + M_3B	1,481	23.2	–	
FeMnNiCo	fcc	1,750	51	129	[23]
(FeMnNiCo)$_{95}$Mo$_5$	fcc	1,800	51	137	
(FeMnNiCo)$_{90}$Mo$_{10}$	fcc + μ	2,250	51	185	
FeCoNiCuCr	fcc1 + fcc2	–	–	–	[24]
(FeCoNiCuCr)$_{96}$Nb$_4$	fcc1 + fcc2 + laves	–	–	–	
(FeCoNiCuCr)$_{90}$Nb$_{10}$	fcc1 + laves	–	–	–	
Mo$_{0.25}$V$_{0.25}$Ti$_{1.5}$Zr$_{0.5}$	bcc	1244.25	9.88	408.6	[25]
Mo$_{0.25}$V$_{0.25}$Ti$_{1.5}$Zr$_{0.5}$Nb$_{0.25}$	bcc	1473.81	19.57	387.2	
Mo$_{0.25}$V$_{0.25}$Ti$_{1.5}$Zr$_{0.5}$Nb$_{0.5}$	bcc	1470.75	22.23	346.6	
Mo$_{0.25}$V$_{0.25}$Ti$_{1.5}$Zr$_{0.5}$Nb$_{0.75}$	bcc	1416.54	23.63	354.8	
Mo$_{0.25}$V$_{0.25}$Ti$_{1.5}$Zr$_{0.5}$Nb	bcc	1307.26	28.32	349.9	
TiZrHf	HCP	1,184	17	211	[26]
TiZrHfSc	HCP	1,800	21.9	233	
TiZrHfY	HCP	1,071	17.7	241	
TiZrHfScY	HCP	1,365	15.7	256	
NbTaWMo	bcc	1,499	3.8	504.5	[27]
NbTaWMoSi$_{0.25}$	bcc+(Nb, Ta)$_5$Si$_3$	2,548	10.5	567	
NbTaWMoSi$_{0.5}$	bcc+(Nb, Ta)$_5$Si$_3$	2,454	5.8	697	
NbTaWMoSi$_{0.75}$	bcc+(Nb, Ta)$_5$Si$_3$	2,732	1.6	682.6	
FeCoCrNi	fcc	–	–	89.52	[28]
FeCoCrNiSi	fcc + bcc	–	–	653.71	

Table 8.1 (continued)

HEAs	Phase/ structure	Compressive strength (MPa)	Compressive strain (%)	Hardness (HV)	References
$FeCoNi_{1.5}CuB$	fcc	1,456	–	–	[29]
$FeCoNi_{1.5}CuBY_{0.1}$	fcc + M_3B	1,602	–	–	
$FeCoNi_{1.5}CuBY_{0.2}$	fcc + M_3B	1,746	–	–	
$FeCoNi_{1.5}CuBY_{0.5}$	fcc + M_3B	1,556	–	–	
(CuCoFeNi)	Fcc	–	–	157.3	[30]
$(CuCoFeNi)Ti_{0.2}$	Fcc	–	–	223.3	
$(CuCoFeNi)Ti_{0.4}$	fcc	1226.8	54.9	389.2	
$(CuCoFeNi)Ti_{0.6}$	fcc + laves	1652.3	34.1	453.3	
$(CuCoFeNi)Ti_{0.8}$	fcc + laves	1723.4	27.9	461.6	
(CuCoFeNi)Ti	fcc + laves	1451.5	18.3	483.0	
TiZrVNb	bcc + V_2Zr	2,287	35.2	434.9	[30]
$Ti_{1.5}ZrVNb$	bcc	–	–	383.1	
Ti_2ZrVNb	bcc	–	–	350.4	
$CoFeNiMnV_{0.25}$	fcc1 + fcc2	–	–	213	[31]
$CoFeNiMnV_{1.25}$	fcc + σ	1,545	26		
$CoFeNiMnV_{1.5}$	fcc + σ	1,678	9.5	716	
FeCrNiMnCo	fcc + bcc	–	–	145	[32]
$FeCrNiMnCoZr_{x0.1}$	fcc + bcc	–	–	174	
$FeCrNiMnCoZr_{x0.2}$	fcc + bcc	–	–	258	
$FeCrNiMnCoZr_{x0.3}$	fcc + bcc	–	–	394	
$Al_{0.5}FeCrNiMnCo$	fcc + bcc	–	–	270	
$Al_{0.5}FeCrNiMnCoZr_{0.1}$	fcc + bcc	–	–	363	
$Al_{0.5}FeCrNiMnCoZr_{0.2}$	fcc + bcc	–	–	449	
$Al_{0.5}FeCrNiMnCoZr_{0.3}$	fcc + bcc	–	–	519	

Note: HCP: hexagonal close-packed; **BCC**: body-centered cubic; **FCC**: face-centered cubic.

8.3 Conclusions

Effect of alloying elements on the phases of various HEA systems were discussed and analyzed in this chapter. Several alloying elements namely Co, Nb, Cu, and Mo were found to promote the formation of FCC phase, while Al, Ti, and Mo alloying elements tend to enhance the formation of BCC phase in HEAs. It is noticeable that Mo acted as BCC as well as FCC, subject to conditions. Moreover, there are very few but impactful literature reporting the improvement in both strength and ductility by incorporating an alloying element like Mo. Therefore, it is recommended to explore the compositional space for such potential HEAs that would deliver the high-performing alloys for mass production with desired properties. This could be achieved by optimizing the composition of HEAs prior to development, with the help of statistical tools or by heuristic search algorithms, such as genetic algorithms.

References

[1]   Z. Rong, et al., "Microstructure and properties of FeCoNiCrX (X Mn, Al) high-entropy alloy coatings," *Journal of Alloys and Compounds*, vol. 921, p. 166061, Nov. 2022, doi: 10.1016/j.jallcom.2022.166061.

[2]   X. Li, et al., "Microstructure evolution and mechanical properties of AlCrFe2Ni2(MoNb)x high entropy alloys," *Journal of Materials Research and Technology*, vol. 17, pp. 865–875, Mar. 2022, doi: 10.1016/j.jmrt.2022.01.055.

[3]   A. K. Sinha, V. K. Soni, R. Chandrakar, and A. Kumar, "Influence of refractory elements on mechanical properties of high entropy alloys," *Transactions of the Indian Institute of Metals*, vol. 74, no. 12, Springer, pp. 2953–2966, Dec. 01. 2021, doi: 10.1007/s12666-021-02363-x.

[4]   S. Yamanaka, K. Ichi Ikeda, and S. Miura, "The effect of titanium and silicon addition on phase equilibrium and mechanical properties of CoCrFeMnNi-based high entropy alloy," *Journal of Materials Research*, vol. 36, no. 10, pp. 2056–2070, May 2021, doi: 10.1557/s43578-021-00251-0.

[5]   A. Yakın, T. Şimşek, B. Avar, A. K. Chattopadhyay, S. Özcan, and T. Şimşek, "The effect of Cr and Nb addition on the structural, morphological, and magnetic properties of the mechanically alloyed high entropy FeCoNi alloys," *Applied Physics A*, vol. 128, no. 8, pp. 686, Aug 2022, doi: 10.1007/s00339-022-05836-y.

[6]   Q. Shen, J. Xue, X. Yu, Z. Zheng, and N. Ou, "Powder plasma arc additive manufacturing of CoCrFeNiWx high-entropy alloys: Microstructure evolution and mechanical properties," *Journal of Alloys and Compounds*, vol. 922, p. 166245, Nov. 2022, doi: 10.1016/j.jallcom.2022.166245.

[7]   C. Y. Hsu, W. R. Wang, W. Y. Tang, S. K. Chen, and J. W. Yeh, "Microstructure and mechanical properties of new AlCoxCrFeMo 0.5Ni high-entropy alloys," *Advanced Engineering Materials*, vol. 12, no. 1–2, pp. 44–49, Feb 2010, doi: 10.1002/adem.200900171.

[8]   J. Hu, Y. Liu, and W. Li, "Microstructure and corrosion behavior of FeCrNiMnMox high-entropy alloys fabricated by the laser surface remelting," *Materials and Corrosion*, no. April, pp. 1–8, 2020, doi: 10.1002/maco.202011674.

[9]   R. Li, et al., "Novel (CoFe2NiV0.5Mo0.2)100−xNbx Eutectic High-Entropy Alloys with Excellent Combination of Mechanical and Corrosion Properties," *Acta MetallurgicaSinica (English Letters)*, 2020, doi: 10.1007/s40195-020-01072-6.

[10] L. Huang, X. Wang, B. Huang, X. Zhao, H. Chen, and C. Wang, "Effect of Cu segregation on the phase transformation and properties of AlCrFeNiTiCux high-entropy alloys," *Intermetallics (Barking)*, vol. 140, Jan. 2022, doi: 10.1016/j.intermet.2021.107397.

[11] V. K. Soni, S. Sanyal, and S. K. Sinha, "Phase evolution and mechanical properties of novel FeCoNiCuMox high entropy alloys," *Vacuum*, vol. 174, no. December, 2019, p. 109173, 2020, doi: 10.1016/j.vacuum.2020.109173.

[12] C. Y. Hsu, C. C. Juan, T. S. Sheu, S. K. Chen, and J. W. Yeh, "Effect of aluminum content on microstructure and mechanical properties of Alx CoCrFeMo0.5Ni high-entropy alloys," *JOM*, vol. 65, no. 12, pp. 1840–1847, Dec 2013, doi: 10.1007/s11837-013-0753-6.

[13] M. Zhang, X. Zhou, and J. Li, "Microstructure and mechanical properties of a refractory CoCrMoNbTi high-entropy alloy," *Journal of Materials Engineering and Performance*, vol. 26, no. 8, pp. 3657–3665, 2017, doi: 10.1007/s11665-017-2799-z.

[14] K. R. Rao and S. K. Sinha, "Effect of sintering temperature on microstructural and mechanical properties of SPS processed CoCrCuFeNi based ODS high entropy alloy," *Materials Chemistry and Physics*, vol. 256, Dec. 2020, doi: 10.1016/j.matchemphys.2020.123709.

[15] K. Raja Rao and S. K. Sinha, "Strengthening of AlCoCrFeNi based high entropy alloy with nano- Y2O3 dispersion," *Materials Science and Engineering B: Solid-State Materials for Advanced Technology*, vol. 281, Jul. 2022, doi: 10.1016/j.mseb.2022.115720.

[16] R. Chandrakar, A. Kumar, S. Chandraker, K. R. Rao, and M. Chopkar, "Microstructural and mechanical properties of AlCoCrCuFeNiSix (x = 0 and 0.9) high entropy alloys," *Vacuum*, vol. 184, Feb. 2021, doi: 10.1016/j.vacuum.2020.109943.

[17] K. Raja Rao and S. K. Sinha, "Phase evolution in novel Y2O3 dispersed CrCuFeNiZn nanocrystalline multicomponent alloy prepared by mechanical alloying," *Vacuum*, vol. 184, Feb. 2021, doi: 10.1016/j.vacuum.2020.109802.

[18] K. R. Rao and S. K. Sinha, "Synthesis and Phase Investigation of Equiatomic AlCrFeMnNi Alloys Dispersed with Partially Stabilized Zirconia for Nuclear Applications," 2020. [Online]. Available: www.scientific.net.

[19] H. Wang, K. Lu, S. Fan, Y. Liu, Y. Zhao, and F. Yin, "Effect of Al content on the microstructure and properties of CoCrCuFeNiMoAlx high entropy alloy," *Materials Today Communications*, vol. 32, p. 103918, Aug. 2022, doi: 10.1016/j.mtcomm.2022.103918.

[20] K. Lu, et al., "Effect of Al elements on the microstructure and properties of CoCrNiCuMoAlx high-entropy alloys," *JOM*, 2022, doi: 10.1007/s11837-022-05267-3.

[21] Y. Li, Z. Yang, Z. Ma, Y. Bai, C. Wu, and J. Li, "Effect of element V on the as-cast microstructure and mechanical properties of Al0.4Co0.5VxFeNi high entropy alloys," *Journal of Alloys and Compounds*, vol. 911, Aug. 2022, doi: 10.1016/j.jallcom.2022.165043.

[22] H. Zhang, et al., "Effect of high configuration entropy and rare earth addition on boride precipitation and mechanical properties of multi-principal-element alloys," *Journal of Materials Engineering and Performance*, vol. 26, no. 8, pp. 3750–3755, Aug. 2017, doi: 10.1007/s11665-017-2831-3.

[23] K. Cichocki, P. Bała, T. Kozieł, G. Cios, N. Schell, and K. Muszka, "Effect of Mo on phase stability and properties in FeMnNiCo high-entropy alloys," *Metallurgical and Materials Transactions A: Physical Metallurgy and Materials Science*, vol. 53, no. 5, pp. 1749–1760, May 2022, doi: 10.1007/s11661-022-06629-x.

[24] A. Y. Churyumov, A. V. Pozdniakov, A. I. Bazlov, H. Mao, V. I. Polkin, and D. V. Louzguine-Luzgin, "Effect of Nb addition on microstructure and thermal and mechanical properties of Fe-Co-Ni-Cu-Cr multiprincipal-element (high-entropy) alloys in as-cast and heat-treated state," *JOM*, vol. 71, no. 10, pp. 3481–3489, Oct 2019, doi: 10.1007/s11837-019-03644-z.

[25] F. Zhang, C. Xiang, E. H. Han, and Z. Zhang, "Effect of Nb content on microstructure and mechanical properties of Mo0.25V0.25Ti1.5Zr0.5Nbx high-entropy alloys," *Acta MetallurgicaSinica (English Letters)*, 2022, doi: 10.1007/s40195-022-01399-2.

[26] T. Huang, H. Jiang, Y. Lu, T. Wang, and T. Li, "Effect of Sc and Y addition on the microstructure and properties of HCP-structured high-entropy alloys," *Applied Physics A: Materials Science and Processing*, vol. 125, no. 3, Mar. 2019, doi: 10.1007/s00339-019-2484-1.

[27] Z. Guo, A. Zhang, J. Han, and J. Meng, "Effect of Si additions on microstructure and mechanical properties of refractory NbTaWMo high-entropy alloys," *Journal of Materials Science*, vol. 54, no. 7, pp. 5844–5851, Apr 2019, doi: 10.1007/s10853-018-03280-z.

[28] L. Huang, et al., "Effect of Si element on phase transformation and mechanical properties for FeCoCrNiSix high entropy alloys," *Materials Letters*, vol. 282, Jan. 2021, doi: 10.1016/j.matlet.2020.128809.

[29] G. R. Li, et al., "Effect of the rare earth element yttrium on the structure and properties of boron-containing high-entropy alloy," *JOM*, vol. 72, no. 6, pp. 2332–2339, Jun. 2020, doi: 10.1007/s11837-020-04059-x.

[30] X. Cong Ye, et al., "Effect of Ti content on microstructure and mechanical properties of CuCoFeNi high-entropy alloys," *International Journal of Minerals, Metallurgy and Materials*, vol. 27, no. 10, pp. 1326–1331, Oct. 2020, doi: 10.1007/s12613-020-2024-1.

[31] M. Zhu, et al., "Effect of V content on phase formation and mechanical properties of the CoFeNiMnVx high-entropy alloys," *Journal of Materials Engineering and Performance*, vol. 31, no. 4, pp. 3151–3158, Apr. 2022, doi: 10.1007/s11665-021-06428-2.

[32] S. S. M. Pauzi, W. Darham, R. Ramli, M. K. Harun, and M. K. Talari, "Effect of Zr addition on microstructure and properties of FeCrNiMnCoZr x and Al0.5FeCrNiMnCoZr x high entropy alloys," *Transactions of the Indian Institute of Metals*, vol. 66, no. 4, pp. 305–308, Aug 2013, doi: 10.1007/s12666-013-0264-8.

Agnivesh Kumar Sinha, Harendra Kumar Narang,
Somnath Bhattacharya, Chandan Pandey

Chapter 9
Effect of alloying elements on the properties of high-entropy alloys

Abstract: High-entropy alloys (HEAs) consist of five or more major alloying elements, resulting in outstanding characteristics, namely, compressive strength, high temperature strength, and hardness. For a wide range of applications, specific properties such as, compressive strength and strain, tensile strength and strain, and hardness, or their combinations are necessary. However, it also noticeable that achieving a balance of properties (like strength and ductility) in HEAs is still a big hurdle for its wide application. Therefore, this chapter emphasizes on the influence of alloying elements on the properties of HEAs, which would enable researchers to design HEAs with the desired attributes for specific applications.

Keywords: High-entropy alloys, Compressive properties, Hardness, Tensile properties

9.1 Introduction

A high-entropy alloy (HEA) usually constitutes five or more major alloying elements in **atomic ratios** ranging from 5% to 35% for each of the individual elements [1]. This results in an increase in the entropy of mixing. Unlike a conventional alloy system, HEAs do not have any main alloying element that majorly governs the properties of the alloy. In HEAs, the properties of alloy depend on all the major constituent elements. In general, HEAs have excellent wear resistance, corrosion resistance, strength, oxidation resistance, hardness, antibacterial properties, etc. For this reason, HEAs have become quite popular and have attracted many researchers [2] in the last decade. These attributes of HEAs are the result of the formation of a **solid solution,** which in turn manifests into solid solution strengthening and grain boundary strengthening.

Researchers have investigated and analyzed [3] the influence of Al on corrosion and on the structure of $Al_xCuFeNiCoCr$ HEAs, wherein Al was varied from 0.5 to 2 atomic ratio in order to understand its effect on the corrosion resistance and microstructure of

Agnivesh Kumar Sinha, Mechanical Engineering Department, Rungta College of Engineering and Technology, Bhilai, Chhattisgarh 490024, India, e-mail: sinhaagnivesh@yahoo.in
Harendra Kumar Narang, Somnath Bhattacharya, Mechanical Engineering Department, National Institute of Technology, Raipur, Chhattisgarh 492010, India
Chandan Pandey, Department of Mechanical Engineering, Indian Institute of Technology Jodhpur, Karwar, Rajasthan 342037, India

https://doi.org/10.1515/9783110769470-009

HEAs. It was observed from the results that the presence of Al promoted a BCC structure, which in turn improves the **electrochemical corrosion** characteristics of HEAs. Likewise, the influence of Co on the tensile properties of HEAs was also investigated [4]. Here, the introduction of Co into $AlCo_xCrFeNi$ enhanced its plasticity without affecting its strength. However, another research on $AlCoCrFe_xNi$ showed that the addition of Fe composition in HEA improved its plasticity at the cost of compressive strength [5]. Results also showed a deterioration in the hardness of HEAs. The deterioration in the properties of HEAs occurred due to the addition of Fe content, which resulted in decreasing the lattice distortion of change of the structure. Replacing Mn with Ge in CoCrFeMnNi HEA improved its tensile characteristics by a significant margin, which was attributed to enhanced elastic moduli and improved bonding, which in turn resulted in the formation of its twin in the vicinity of plastic regions [6].

Past literature showed that a variation in the composition of the alloying element has a direct or indirect effect on the properties of HEAs. This may be due to the transformation in the phase/structure of HEA or due to the change in its strengthening mechanism. Thus, it is of vital importance to explore the probable space for an HEA system for the various possible combinations of its compositions. Moreover, to design or develop a suitable HEA for a specific application, it is required to understand the effect of different alloying elements on the properties of HEAs, namely tensile properties, compressive properties, and hardness.

9.2 Effect of alloying element on the properties of HEAs

The influence of alloying elements along with their composition on properties such as compressive strength, **compressive strain**, tensile strength, tensile strain, and hardness is an important aspect that has been further discussed in this chapter.

9.2.1 Compressive strength and strain

Compressive strength is the key criterion for any load-bearing structure or component that is under compressive load condition. Thus, researchers have investigated the effect of Al composition on the compressive strength of $CoCrCuFeNiMoAl_x$ HEAs [7]. Results showed improvement in the compressive strength of HEA of up to 22% due to the increase in Al content, resulting in the formation of a BCC structure. In a similar study, the effect of Al was on the mechanical properties $CoCrNiCuMoAl_x$ was analyzed, where Al was varied from $x = 0$ to $x = 1.0$ [8]. Results showed that the compressive strength of HEA greatly improved by increasing the Al content. A compressive strength of 1,775 MPa was achieved by the addition of Al in $CoCrNiCuMoAl_x$ at x = 1. CoCrNiCuMoAl showed an increment of almost 21% when compared to the CoCrNiCuMo HEA. An increase in compressive strength was observed due to the transformation of the FCC structure into a BCC structure by the incorporation of Al in HEAs. This also promoted the formation of solid solution, which in turn improved the strength.

Addition of Yttrium in CoCrFeNi-B resulted in increasing the compressive strength by approximately 50% (for $CoCrFeNiY_{0.1}$-B) [9]. Moreover, the compressive strain improved from 4.9% (CoCrFeNi-B) to 12.5% ($CoCrFeNiY_{0.1}$-B). Interstitial solid solution strengthening was attributed to the improvement in compressive strength and strain of $CoCrFeNiY_{0.1}$-B. Similar improvement was also observed in the same study for $CoMnFeNiAl_{0.3}Cu_{0.7}$-B HEA, wherein the incorporation of Yttrium resulted in increasing the compressive strain and stress up to 23.2% and 1,481 MPa, respectively for $CoMnFeNiAl_{0.3}Cu_{0.7}Y_{0.1}$-B HEA. It was also reported that the addition of V content in HEA enhanced the compressive strength up to 3,000 MPa but at the expense of elongation [10]. FeCoNiCuCr HEA also exhibited an improvement of about 100% in compressive strength on the introduction of Nb [11]. A study reported a deterioration in the compressive strength of HEA with the addition of Y content and an improvement in the compressive strength by the addition of Sc [12]. The decrease in the compressive strength of HEA due to the incorporation of Y was attributed to the formation of the electride structure (α-like) whereas the increase in the compressive strength of HEA was due to the HCP structure. However, in another research [13], it was shown that on incorporating Y content in $FeCoNi_{1.5}CuBY_x$ HEAs (x = 0, 0.1,0.2,0.5), the compressive strength first increases up to x = 0.2 and then decreases at x = 0.5. This showed that 0.2 atomic wt% of Y was optimum for the HEA attaining the maximum compressive strength.

9.2.2 Tensile strength and strain

Tensile properties of any component are vital for any structural application. Tensile strain and stress are majorly governed by the alloying elements of HEAs. Research has shown that various alloying elements influence tensile properties in different ways.

The tensile strength of $Al_{0.4}Co_{0.5}V_xFeNi$ HEA increases abruptly on adding V from x = 0 to x = 0.4 [10]. Moreover, tensile strain also increases in excess of 30% due to the incorporation of V content in HEA. However, on increasing the V content in HEA to x = 0.6, the tensile strength further increases but at the expense of tensile strain. Thus, x = 0.4 is more suitable for applications that require a balance of strength and ductility. The influence of Mo content on the tensile strength of $(FeMnNiCo)_{100-x}Mo_x$ HEAs was determined in a research, which showed that the strength increased significantly with the addition of Mo [14]. Interestingly, it was observed in research that the incorporation of Ti content in TiZrVNb HEA resulted in a decrease in the tensile strength. The tensile strength for $Ti_{1.5}ZrVNb$ and Ti_2ZrVNb was reported as 1,401 MPa and 1,305 MPa, respectively. This shows that the tensile strength of $Ti_{1.5}ZrVNb$ and Ti_2ZrVNb was reduced by approximately 3% and 9%, respectively. Hence, Ti addition is not suggested for improving the

tensile strength of HEA beyond 1 atomic wt%. From this result, it can be inferred that deter-
mination of the optimum composition is necessary for developing HEAs for the desired
applications.

9.2.3 Hardness

Hardness of HEA is an important aspect for its application, as it represents the toughness, which is an important criterion. Thus, hardness, specifically micro-hardness, of
HEAs matters a lot before deciding their application. Literature shows that the hardness of HEAs is greatly affected by its alloying elements.

Hardness increases with the increase in the V composition in $Al_{0.4}Co_{0.5}V_xFeNi$ HEA
when it is varied from 0 to 0.6 [10]. Hardness was also found to increase on the addition
of Mo from $x = 0$ to $x = 10$, in steps of 5 atomic wt%, for $(FeMnNiCo)_{100-x}Mo_x$ HEAs [14].
Results showed that this improvement in the hardness of HEAs was due to the solid
solution strengthening, grain boundary strengthening and precipitation strengthening.
There was an increase of about 43% in HEAs having 5 atomic wt% ($x = 5$) when compared to an HEA without Mo ($x = 0$). While, in another study, it was reported that on
increasing the Nb content in $Mo_{0.25}V_{0.25}Ti_{1.5}Zr_{0.5}Nb_x$HEA, the hardness tends to deterio-
rate [15]. In another very interesting study [12], it was observed that the addition of Sc
and Y in TiZrHf HEA improved its hardness but it was noteworthy that the incorporation of Y in the HEA improved its hardness more than that by the addition of Sc.
Moreover, results also showed that the incorporation of both Sc and Y in HEA enhanced
the hardness of HEA more than that when Sc and Y is incorporated individually.

It is observed in research [16] that the addition of Si in $NbTaWMoSi_x$ HEAs also resulted in improving the hardness when its composition is varied, that is at $x = 0$, 0.25,
0.5, 0.75. The improvement in its hardness was due to the formation of the BCC and sili-
cide structure. In another research on $FeCoCrNiSi_x$, HEAs [17] showed that when Si is
varied from 0 to 1 in steps of 0.5 atomic wt%, the hardness tends to increase. The hardness of HEA increased by 600%, which was attributed to the NiSi phase formation, and
an increase in the lattice distortion energy due to the addition of Si with its small atomic
size. Table 9.1 tabulates the various mechanical properties of HEAs.

9.3 Conclusions and future prospects

This chapter dealt with the effects of HEA alloying elements on their properties,
namely compressive stress and strain, tensile stress and strain, and hardness. It is distinct from the literature survey that many researches have conducted for investigat-
ing the influence of alloying on the mechanical properties of HEAs, but there is a
scarcity of literature relevant to the tensile properties of HEAs. This might be due to

Table 9.1: Mechanical properties of HEAs.

HEAs	Tensile strength (MPa)	Tensile strain	Compressive strength (MPa)	Compressive strain	Hardness (HV)	References
CoCrCuFeNiMo	–	–	1,624	–	475.09	[7]
CoCrCuFeNiMoAl$_{0.32}$	–	–	1,486	–	521.75	
CoCrCuFeNiMoAl$_{0.67}$	–	–	1646.5	–	535.37	
CoCrCuFeNiMoAl	–	–	1,826	–	619.73	
CoCrNiCuMo	–	–	1,463	–	540.5	[8]
CoCrNiCuMoAl$_{0.26}$	–	–	1,735	–	572.6	
CoCrNiCuMoAl$_{0.56}$	–	–	1,750	–	613.4	
CoCrNiCuMoAl	–	–	1,775	–	621.5	
AlCrFeNiTi	–	–	–	–	434	[18]
AlCrFeNiTiCu$_{0.5}$	–	–	–	–	382	
AlCrFeNiTiCu	–	–	–	–	358	
AlCrFeNiTiCu$_{1.5}$	–	–	–	–	320	
AlCrFeNiTiCu$_2$	–	–	–	–	280	
Al$_{0.4}$Co$_{0.5}$FeNi	430	70.1	2,000	50	–	[10]
Al$_{0.4}$Co$_{0.5}$V$_{0.2}$FeNi	722	34.1	3,000	60	–	
Al$_{0.4}$Co$_{0.5}$V$_{0.4}$FeNi	809	33.5	3,000	60	–	
Al$_{0.4}$Co$_{0.5}$V$_{0.6}$FeNi	1,128	27.7	1,746	36.4	–	
CoCrFeNiB	–	–	895	4.9	–	[9]
CoCrFeNiY$_{0.1}$B	–	–	1,350	15	–	
CoMnFeNiAl$_{0.3}$Cu$_{0.7}$B	–	–	1,205	19.4	–	
CoMnFe-NiAl$_{0.3}$Cu$_{0.7}$Y$_{0.1}$B	–	–	1,481	23.2	–	
FeMnNiCo	671	36	1,750	51	129	[14]
(FeMnNiCo)$_{95}$Mo$_5$	891	34	1,800	51	137	
(FeMnNiCo)$_{90}$Mo$_{10}$	1,070	12	2,250	51	185	
FeCoNiCuCr	520	70	–	–	–	[19]
(FeCoNiCuCr)$_{96}$Nb$_4$	1,010	52	–	–	–	
(FeCoNiCuCr)$_{90}$Nb$_{10}$	1,495	17	–	–	–	

Table 9.1 (continued)

HEAs	Tensile strength (MPa)	Tensile strain	Compressive strength (MPa)	Compressive strain	Hardness (HV)	References
$Mo_{0.25}V_{0.25}Ti_{1.5}Zr_{0.5}$	–	–	1244.25	9.88	408.6	[15]
$Mo_{0.25}V_{0.25}Ti_{1.5}Zr_{0.5}Nb_{0.25}$	–	–	1473.81	19.57	387.2	
$Mo_{0.25}V_{0.25}Ti_{1.5}Zr_{0.5}Nb_{0.5}$	–	–	1470.75	22.23	346.6	
$Mo_{0.25}V_{0.25}Ti_{1.5}Zr_{0.5}Nb_{0.75}$	–	–	1416.54	23.63	354.8	
$Mo_{0.25}V_{0.25}Ti_{1.5}Zr_{0.5}Nb$	–	–	1307.26	28.32	349.9	
TiZrHf	–	–	1,184	17	211	[12]
TiZrHfSc	–	–	1,800	21.9	233	
TiZrHfY	–	–	1,071	17.7	241	
TiZrHfScY	–	–	1,365	15.7	256	
NbTaWMo	–	–	1,499	3.8	504.5	[16]
$NbTaWMoSi_{0.25}$	–	–	2,548	10.5	567	
$NbTaWMoSi_{0.5}$	–	–	2,454	5.8	697	
$NbTaWMoSi_{0.75}$	–	–	2,732	1.6	682.6	
FeCoCrNi	–	–	–	–	89.52	[17]
FeCoCrNiSi	–	–	–	–	653.71	
$FeCoNi_{1.5}CuB$	–	–	1,456	–	–	[13]
$FeCoNi_{1.5}CuBY_{0.1}$	–	–	1,602	–	–	
$FeCoNi_{1.5}CuBY_{0.2}$	–	–	1,746	–	–	
$FeCoNi_{1.5}CuBY_{0.5}$	–	–	1,556	–	–	
(CuCoFeNi)	–	–	–	–	157.3	[20]
$(CuCoFeNi)Ti_{0.2}$	–	–	–	–	223.3	
$(CuCoFeNi)Ti_{0.4}$	–	–	1226.8	54.9	389.2	
$(CuCoFeNi)Ti_{0.6}$	–	–	1652.3	34.1	453.3	
$(CuCoFeNi)Ti_{0.8}$	–	–	1723.4	27.9	461.6	
(CuCoFeNi)Ti	–	–	1451.5	18.3	483.0	
TiZrVNb	1,447	3.5	2,287	35.2	434.9	[21]
$Ti_{1.5}ZrVNb$	1,401	11.2	–	–	383.1	
Ti_2ZrVNb	1,305	12.3	–	–	350.4	

Table 9.1 (continued)

HEAs	Tensile strength (MPa)	Tensile strain	Compressive strength (MPa)	Compressive strain	Hardness (HV)	References
CoFeNiMnV$_{0.25}$	–	–	–	–	213	[22]
CoFeNiMnV$_{1.25}$	–	–	1,545	26		
CoFeNiMnV$_{1.5}$	–	–	1,678	9.5	716	
FeCrNiMnCo	–	–	–	–	145	[23]
FeCrNiMnCoZr$_{x0.1}$	–	–	–	–	174	
FeCrNiMnCoZr$_{x0.2}$	–	–	–	–	258	
FeCrNiMnCoZr$_{x0.3}$	–	–	–	–	394	
Al$_{0.5}$FeCrNiMnCo	–	–	–	–	270	
Al$_{0.5}$FeCrNiMnCoZr$_{0.1}$	–	–	–	–	363	
Al$_{0.5}$FeCrNiMnCoZr$_{0.2}$	–	–	–	–	449	
Al$_{0.5}$FeCrNiMnCoZr$_{0.3}$	–	–	–	–	519	
CoCrCuFeNi + 1 wt Y$_2$O$_3$	–	–	1,555	–	468 ± 5	[24]
AlCoCrFeNi	–	–	1613 ± 22	–	581 ± 6	[25]
AlCoCrCuFeNi	–	–	–	–	650	[26]

Note: HCP: hexagonal close-packed; **bcc**: body-centered cubic; **fcc**: face-centered cubic.

the fact that HEAs are mostly intended for applications that require excellent compressive strength. However, to explore more possibilities of application for HEAs, it is suggested to investigate potential alloying elements for developing HEAs with excellent tensile properties too. Moreover, it is also recommended to develop HEAs with a balance of both strength and ductility. For this purpose, it is suggested to optimize the composition of the HEA alloying elements.

Optimization of the composition in HEAs can be achieved by using statistical tools, fuzzy logic, and genetic algorithms. This would not only save time but also enable researchers to explore the huge space of possibilities for designing and developing high-performing HEAs for a broad range of applications.

References

[1] J. W. Yeh, et al., "Nanostructured high-entropy alloys with multiple principal elements: Novel alloy design concepts and outcomes," *Advanced Engineering Materials*, vol. 6, no. 5, pp. 299–303, 2004, doi: 10.1002/adem.200300567.

[2] A. K. Sinha, V. K. Soni, R. Chandrakar, and A. Kumar, "Influence of refractory elements on mechanical properties of high entropy alloys," *Transactions of the Indian Institute of Metals*, vol. 74, no. 12, Springer, pp. 2953–2966, Dec. 01, 2021, doi: 10.1007/s12666-021-02363-x.

[3] R. Hu, J. Du, Y. Zhang, Q. Ji, R. Zhang, and J. Chen, "Microstructure and corrosion properties of AlxCuFeNiCoCr(x=0.5, 1.0, 1.5, 2.0) high entropy alloys with Al content," *Journal of Alloys and Compounds*, p. 165455, Nov. 2022, doi: 10.1016/j.jallcom.2022.165455.

[4] S. Zhao, et al., "Microstructure and enhanced tensile properties of AlCoxCrFeNi high entropy alloys with high Co content fabricated by laser melting deposition," *Journal of Alloys and Compounds*, vol. 917, Oct. 2022, doi: 10.1016/j.jallcom.2022.165403.

[5] Q. Chen, K. Zhou, L. Jiang, Y. Lu, and T. Li, "Effects of Fe content on microstructures and properties of AlCoCrFexNi high-entropy alloys," *Arabian Journal for Science and Engineering*, vol. 40, no. 12, pp. 3657–3663, Dec. 2015, doi: 10.1007/s13369-015-1784-9.

[6] T. N. Lam, et al., "Element effects of Mn and Ge on the tuning of mechanical properties of high-entropy alloys," *Metallurgical and Materials Transactions A: Physical Metallurgy and Materials Science*, vol. 51, no. 10, pp. 5023–5028, Oct. 2020, doi: 10.1007/s11661-020-05932-9.

[7] H. Wang, K. Lu, S. Fan, Y. Liu, Y. Zhao, and F. Yin, "Effect of Al content on the microstructure and properties of CoCrCuFeNiMoAlx high entropy alloy," *Materials Today Communications*, vol. 32, p. 103918, Aug. 2022, doi: 10.1016/j.mtcomm.2022.103918.

[8] K. Lu, et al., "Effect of Al elements on the microstructure and properties of CoCrNiCuMoAlx high-entropy alloys," *JOM*, 2022, doi: 10.1007/s11837-022-05267-3.

[9] H. Zhang, et al., "Effect of high configuration entropy and rare earth addition on boride precipitation and mechanical properties of multi-principal-element alloys," *Journal of Materials Engineering and Performance*, vol. 26, no. 8, pp. 3750–3755, Aug. 2017, doi: 10.1007/s11665-017-2831-3.

[10] Y. Li, Z. Yang, Z. Ma, Y. Bai, C. Wu, and J. Li, "Effect of element V on the as-cast microstructure and mechanical properties of Al0.4Co0.5VxFeNi high entropy alloys," *Journal of Alloys and Compounds*, vol. 911, Aug. 2022, doi: 10.1016/j.jallcom.2022.165043.

[11] A. Y. Churyumov, A. V. Pozdniakov, A. I. Bazlov, H. Mao, V. I. Polkin, and D. V. Louzguine-Luzgin, "Effect of Nb addition on microstructure and thermal and mechanical properties of Fe-Co-Ni-Cu-Cr multiprincipal-element (high-entropy) alloys in as-cast and heat-treated state," *JOM*, vol. 71, no. 10, pp. 3481–3489, Oct. 2019, doi: 10.1007/s11837-019-03644-z.

[12] T. Huang, H. Jiang, Y. Lu, T. Wang, and T. Li, "Effect of Sc and Y addition on the microstructure and properties of HCP-structured high-entropy alloys," *Applied Physics. A, Materials Science & Processing*, vol. 125, no. 3, Mar. 2019, doi: 10.1007/s00339-019-2484-1.

[13] G. R. Li, et al., "Effect of the rare earth element yttrium on the structure and properties of boron-containing high-entropy alloy," *JOM*, vol. 72, no. 6, pp. 2332–2339, Jun. 2020, doi: 10.1007/s11837-020-04059-x.

[14] K. Cichocki, P. Bała, T. Kozieł, G. Cios, N. Schell, and K. Muszka, "Effect of Mo on phase stability and properties in FeMnNiCo high-entropy alloys," *Metallurgical and Materials Transactions A: Physical Metallurgy and Materials Science*, vol. 53, no. 5, pp. 1749–1760, May. 2022, doi: 10.1007/s11661-022-06629-x.

[15] F. Zhang, C. Xiang, E. H. Han, and Z. Zhang, "Effect of Nb content on microstructure and mechanical properties of Mo0.25V0.25Ti1.5Zr0.5Nbx high-entropy alloys," *Acta Metallurgica Sinica (English Letters)*, 2022, doi: 10.1007/s40195-022-01399-2.

[16] Z. Guo, A. Zhang, J. Han, and J. Meng, "Effect of Si additions on microstructure and mechanical properties of refractory NbTaWMo high-entropy alloys," *Journal of Materials Science*, vol. 54, no. 7, pp. 5844–5851, Apr. 2019, doi: 10.1007/s10853-018-03280-z.

[17] L. Huang, et al., "Effect of Si element on phase transformation and mechanical properties for FeCoCrNiSix high entropy alloys," *Materials Letters*, vol. 282, Jan. 2021, doi: 10.1016/j.matlet.2020.128809.

[18] L. Huang, X. Wang, B. Huang, X. Zhao, H. Chen, and C. Wang, "Effect of Cu segregation on the phase transformation and properties of AlCrFeNiTiCux high-entropy alloys," *Intermetallics (Barking)*, vol. 140, Jan. 2022, doi: 10.1016/j.intermet.2021.107397.

[19] A. Y. Churyumov, A. V. Pozdniakov, A. I. Bazlov, H. Mao, V. I. Polkin, and D. V. Louzguine-Luzgin, "Effect of Nb addition on microstructure and thermal and mechanical properties of Fe-Co-Ni-Cu-Cr Multiprincipal-Element (High-Entropy) alloys in as-cast and heat-treated state," *Jom*, vol. 71, no. 10, pp. 3481–3489, 2019, doi: 10.1007/s11837-019-03644-z.

[20] X. Cong Ye, et al., "Effect of Ti content on microstructure and mechanical properties of CuCoFeNi high-entropy alloys," *International Journal of Minerals, Metallurgy and Materials*, vol. 27, no. 10, pp. 1326–1331, Oct. 2020, doi: 10.1007/s12613-020-2024-1.

[21] T. Dang Huang, S. Yu Wu, S. H. Jiang, Y. Ping Lu, T. Min Wang, and T. Ju Li, "Effect of Ti content on microstructure and properties of TixZrVNb refractory high-entropy alloys," *International Journal of Minerals, Metallurgy and Materials*, vol. 27, no. 10, pp. 1318–1325, Oct. 2020, doi: 10.1007/s12613-020-2040-1.

[22] M. Zhu, et al., "Effect of V content on phase formation and mechanical properties of the CoFeNiMnVx high-entropy alloys," *Journal of Materials Engineering and Performance*, vol. 31, no. 4, pp. 3151–3158, Apr. 2022, doi: 10.1007/s11665-021-06428-2.

[23] S. S. M. Pauzi, W. Darham, R. Ramli, M. K. Harun, and M. K. Talari, "Effect of Zr addition on microstructure and properties of FeCrNiMnCoZr x and Al0.5FeCrNiMnCoZr x high entropy alloys," *Transactions of the Indian Institute of Metals*, vol. 66, no. 4, pp. 305–308, Aug. 2013, doi: 10.1007/s12666-013-0264-8.

[24] K. R. Rao and S. K. Sinha, "Effect of sintering temperature on microstructural and mechanical properties of SPS processed CoCrCuFeNi based ODS high entropy alloy," *Materials Chemistry and Physics*, vol. 256, Dec. 2020, doi: 10.1016/j.matchemphys.2020.123709.

[25] K. Raja Rao and S. K. Sinha, "Strengthening of AlCoCrFeNi based high entropy alloy with nano- Y2O3 dispersion," *Materials Science and Engineering B: Solid-State Materials for Advanced Technology*, vol. 281, Jul. 2022, doi: 10.1016/j.mseb.2022.115720.

[26] R. Chandrakar, A. Kumar, S. Chandraker, K. R. Rao, and M. Chopkar, "Microstructural and mechanical properties of AlCoCrCuFeNiSix (x = 0 and 0.9) high entropy alloys," *Vacuum*, vol. 184, Feb. 2021, doi: 10.1016/j.vacuum.2020.109943.

Om Prakash, Rajesh Kumar, Vikrant Tapas, Anil Kumar, Bojanki Naveen
Chapter 10
Emerging processing routes

Abstract: High-entropy alloys (HEAs) are produced using a number of processing techniques. HEAs have been produced in a variety of materials, including films, dense solid castings, and powder metallurgy components. The three types of processing routes – melting and casting, powder metallurgy, and deposition techniques – can be broadly divided into three classes. In order to create HEAs in the form of rods, bars, and ribbons, melting and casting procedures have been used, along with equilibrium and nonequilibrium cooling rates. The vacuum arc melting, vacuum induction melting, and melt spinning processes are the most widely used melt processing methods. The primary solid-state processing method to create sintered goods has been mechanical alloying (MA), followed by sintering. The surface modification methods utilized to create both thin films and thick layers of HEAs on various substrates include plasma nitriding, cladding, and sputtering. This chapter provides a brief overview of the various synthesis and processing methods used to create HEAs. The processing pathways for equiatomic and nonequiatomic HEAs are comparable.

Keywords: Melting and casting, powder metallurgy, sputtering, cladding, high-entropy alloys

10.1 Introduction

The process used to prepare an HEA is essentially the same as that used to prepare conventional alloys. The three main categories of the preparation technique are melting casting, powder metallurgy, and mechanical alloying. Using a filamentous substance, the manufactured bulk material can be shot or processed into a sheet. To ensure that the formed alloy has a high configurational or mixing entropy, each component keeps its concentration higher. In order to create homogenous alloys and optimize the role of each constituent element to the properties of the HEAs while preserving high mixing

Om Prakash, Department of Mechanical Engineering, Jhada Sirha Government Engineering College, Jagdalpur, Bastar, Chhattisgarh 494001, India, e-mail: omprakash@gecjdp.ac.in
Rajesh Kumar, Department of Mechanical Engineering, CSIT, Durg, Chhattisgarh 491001, India
Vikrant Tapas, Department of Mechanical Engineering, NMDC DAV Polytechnic College Geedam, Dantewada, Chhattisgarh, India
Anil Kumar, Department of Mechanical Engineering, Bhilai Institute of Technology, Durg, Chhattisgarh 491001, India
Bojanki Naveen, Department of Mechanical Engineering, National Institute of Technology Karnataka (NITK), Surathkal, Mangalore 575025, India

https://doi.org/10.1515/9783110769470-010

entropy, it is crucial to understand how to alloy the diverse metal components with their varying properties, metallic structures, and melting temperatures [1] According to their dimensions, HEAs can be categorized into four different types: bulk HEAs (three-dimensional HEAs), high-entropy films and coatings (two-dimensional HEAs), HEA fibers (**one-dimensional HEAs**), and HEA powders (zero-dimensional HEAs) [1]. As a result, the production processes for HEAs with the various morphologies will differ. One of the key elements in choosing the appropriate production process is the geometry of HEAs. For instance, powder metallurgy, induction melting, and arc melting are frequently employed to create bulk HEAs. **HEA coatings** are frequently deposited by laser cladding and magnetic sputtering. This section introduces the production processes for bulk, coatings, fibers, and powders of HEAs. The anticipated microstructure of HEAs product should be taken into account while choosing a manufacturing path. An appropriate approach must be used to prepare a particular structure. To make high-entropy metal glasses, for instance, copper-mold suction casting is frequently utilized; to make **single-crystal HEAs**, Bridgman solidification casting is employed; and to make various HEAs with gradient structures, high gravity casting/co-sputtering can be used.

10.2 Processing routes of bulk HEAs

Melting and solidification procedures are used to create bulk HEAs from a liquid state. Pure solid metals are melted into a liquid condition and then blended uniformly during the melting process. The mixture of the melted liquid elements is then solidified under various cooling settings to produce bulk alloys. Another method is to create large quantities of high-entropy alloys via powder metallurgy.

10.3 Melting method

10.3.1 Arc-melting

Arc melting is the most common melting method used for the production of bulk HEAs. Metals are heated and melted using an arc at a high temperature. Due to the high arc temperature, it is appropriate to make refractory HEAs. Pure metals are first produced in accordance with the intended alloy composition. They are placed in the copper crucible that is cooled by water. High-purity inert gas (Ar) is pumped inside the vacuum chamber for arc initiation and works as the protection gas after the necessary vacuum level reaches. The copper crucible and the arc gun are connected to a high voltage, and when the arc needle comes into contact with the copper crucible, an arc is created. After the arc is initiated, the titanium ingot is first melted while being shielded by the high-purity inert gas (Ar), to **scavenge the atmospheric** oxygen. The HEAs are then

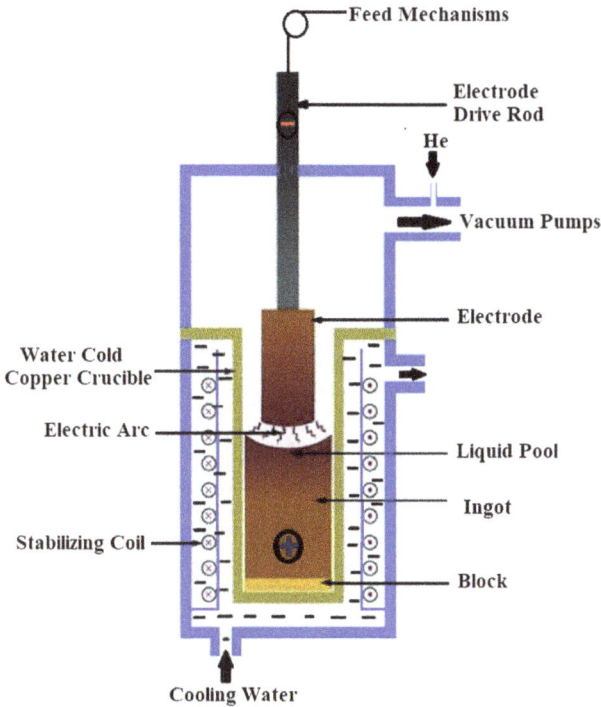

Figure 10.1: Schematic of the vacuum arc melting (VAM) process [2].

produced by melting and solidifying the basic material [2]. The schematic of the vacuum arc melting method is shown in Figure 10.1.

10.4 Induction-levitation-melting (ILM)

Metals often demonstrate strong conductivity, making them efficient and quick to heat, and properly melt metals use induced current. Typically, a copper crucible with an exterior induction coil and water cooling is utilized to generate induced current in the metals. In order to begin the process, the unprocessed, pure metals are first placed in a copper crucible that is located inside a vacuum furnace chamber. A safety gas containing high-quality argon is pumped into the chamber. Induced current is produced in metals after the induction coil is energized. Joule's law states that the induced current in metals produces a significant amount of heat. Induction heating is then used to melt the alloy (Figure 10.2(a)). During the melting process, it is particularly beneficial to homogenize the alloy's composition because of the stirring effect of the induction electromagnetic field. One benefit of this method over arc melting is the capacity to make larger-sized samples. The weight of the ingot can also exceed 1 kg (Figure 10.2(b)) [3].

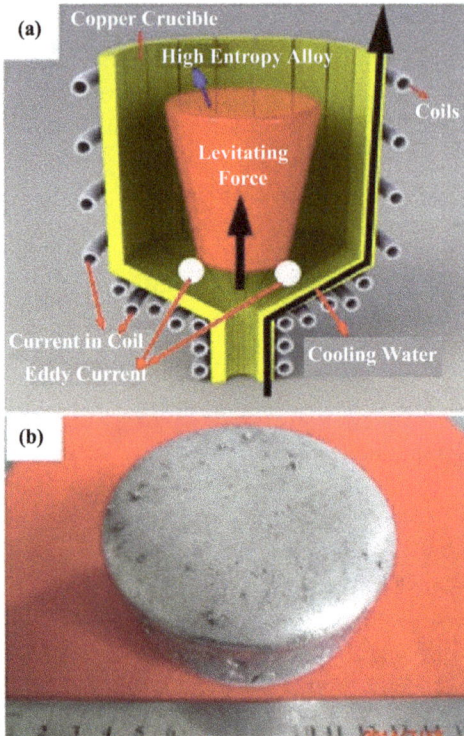

Figure 10.2: (a) Schematic of "Induction-levitation-melting (ILM)"; (b) HEAs sample developed by "induction-levitation-melting" [3].

10.4.1 High-gravity casting

Centrifugal force can be used to replicate the external gravitational field during solidi-fication. During the process, the heavier elements will accumulate along the gravity-dependent axis. As a result, an internal compositional gradient develops in the ingot (Figure 10.3). A series of samples with various compositions are then produced by cut-ting alloy ingots into parts in the direction of gravity [4].

10.5 Additive manufacturing

Although there are numerous techniques to prepare HEA in bulk, there are still signif-icant issues with the preparation methods. Surface coating and surface deposition processes, such as such as laser metal deposition, selective laser melting, are com-monly used in the fabrication of HEAs in additive manufacturing,. It has been difficult

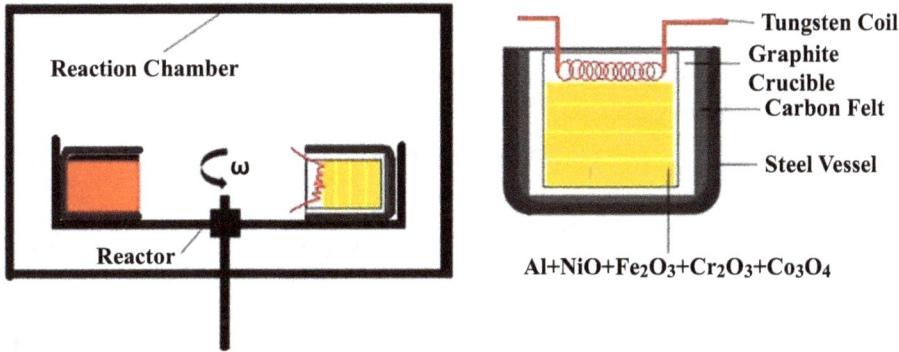

Figure 10.3: The schematic of the "high-gravity combustion synthesis system" [4].

to use refractory-based high-entropy alloys in most large mechanical components because some refractory elements, including Ta and W, have high melting temperatures.
The majority of large-scale bulk HEAs manufactured in the liquid state were developed by ILM.

10.6 Processing method of HEA coatings

10.6.1 Vapor deposition

Vapor deposition technique is the method of depositing various coatings, based on
metal, non-metal, or composite, on the substrate while they are still in the gaseous
phase. Vapor deposition also includes chemical vapor deposition and physical vapor deposition. The three primary categories of physical vapor deposition techniques are ion
coating, sputtering coating, and evaporation coating. The degree of ionization after
atom gasification varies between the three processes, which causes this difference.

The vapor atom ionization rate achieved by evaporation coating is zero, in comparison to the exceptionally high vapor atom ionization rate produced by ion coating.
The most popular sputtering technique is magnetron sputtering, which produces
vapor atoms that are only partially ionized. "Argon atoms are ionized by the electric
field E, producing Ar+ and electrons. Then, when electrons travel to the substrate and
Ar+ ions go to the cathode HEA target, under the acceleration of the electric field, the
target sputters due to the high energy of the impact" [1, 5]. Subsequently, the gasified
ions and atoms form HEA films through deposition on the substrate (Figure 10.4)

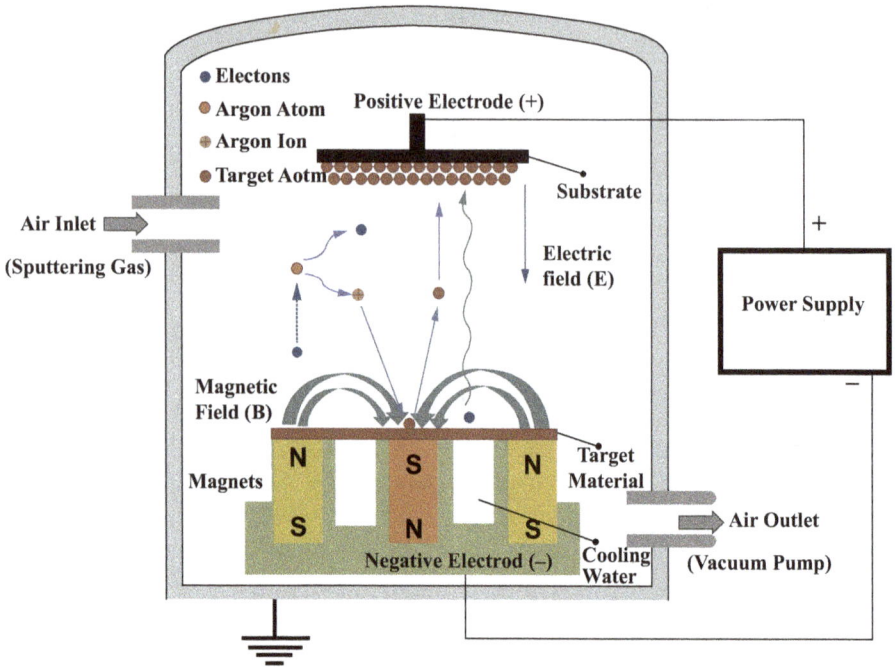

Figure 10.4: Schematic of magnetron sputtering [5].

10.6.2 Surface cladding

In order to create a "metallurgical bond with the substrate surface, the additive clad-ding layer first covers the substrate surface with a cladding material before melting it into a thin layer, using energy heating" [1]. This process is known as surface cladding. It can be separated into laser cladding and plasma cladding, depending on the heating method. In a laser cladding process, a high temperature laser is used for the deposi-tion of melted high-entropy alloy coating on the substrate. Figure 10.5 shows the sche-matic of laser cladding surface coating.

10.7 Electrochemical deposition

In the electroplating process, electrodeposition is used to create pure metal films out of metals like Fe, Ni, Cr, Zn, Cu and some other non-metallic alloys. Large-scale production is currently possible. However, the majority of the ternary and higher alloy electrodepo--sition technology is still in the testing phase. Finding appropriate complexing agents to simultaneously dissolve several ions in a solution is the primary step in the preparation

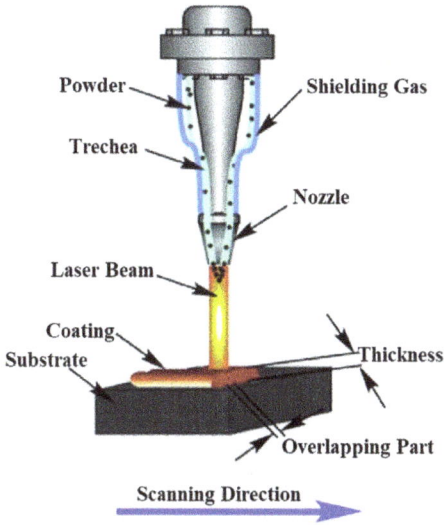

Figure 10.5: Schematic of laser cladding coating process [6].

of HEA coatings by electrodeposition. Following this, metal ions are subsequently deposited into thin layers by cathodic discharge. In general, electrodeposition produces films with densities and interfacial bonding energies that are superior to those made by magnetron sputtering and the further vapor deposition techniques.

The electrodeposition method of TM-Bi and "RE-TM-Bi alloy" films was enhanced in 2008 by Yao et al. [2]. "By dissolving bismuth nitrate and the chloride salts of iron, cobalt, nickel, and manganese in a DMF-acetonitrile-lithium perchlorate organic solution system, Bi-Fe-Co-Ni-Mn HEA films were effectively electrodeposited for the first time" [7].

10.8 Manufacturing routes of HEAs powder

10.8.1 Mechanical alloying

High-speed stirring is applied to the grinding ball in order to repeatedly compress and weld the element powder. The advantages of this method is its low cost and a straightforward approach. The homogeneity of the alloy's constituent parts is facilitated by this process. The alloy powder is continually compressed, crushed, and welded, as the MA is being made. "The powder is continuously subjected to multiple-directional forces, such as compressive force, impact force, and shearing force, during the crushing process." Between the alloy powders, there is a solid-state reaction and diffusion that leads to the formation of nanocrystalline and amorphous structure, and a consistent microstructure [8].

10.9 Carbon-thermal shock method

By using the carbon-thermal shock (CTS) method, immiscible elements are generally alloyed into single-phase nanoparticles on carbon supports with the following characteristics: (i) high mixing entropy; (ii) nonequilibrium processing, in which the shock process creates HEA in milliseconds; and (iii) uniform dispersion. The CTS method's maximum temperature (2,000 to 3,000 K) is generally higher than the temperature at which any metal salt decomposes, promoting consistent mixing of almost any metallic combination [9].

10.10 Future direction

High-entropy alloy's design concept overcomes the limitations of conventional materials. The high thermal stability, high toughness, high strength with plasticity, high magnetic and anti-irradiation, property of the HEAs show considerable application potential. As the research has progressed, HEA composition designs have become more adaptable and varied. According on their histological structure, HEAs have been identified that can form single or multi-phase solid solutions, as well as nano-crystalline, amorphous HEAs. Due to the various components and high alloying elements that offer significant issues to the HEA community, different phase compositions and microstructures affect the overall performance. The effectiveness of the conventional manufacturing process has lagged behind the rate of HEA development.

References

[1] G. M. C. Y., Y. J. W, L. P. K., and Z. Y. Zhang, *High-Entropy Alloys*. Cham: Springer International Publishing, 2016. doi: 10.1007/978-3-319-27013-5.
[2] J. Feng, K. Song, S. Liang, X. Guo, and Y. Jiang, "Electrical wear of TiB2 particle-reinforced Cu and Cu–Cr composites prepared by vacuum arc melting," *Vacuum*, vol. 175, p. 109295, May. 2020, doi: 10.1016/j.vacuum.2020.109295.
[3] S. Xia, M. C. Gao, T. Yang, P. K. Liaw, and Y. Zhang, "Phase stability and microstructures of high entropy alloys ion irradiated to high doses," *Journal of Nuclear Materials*, vol. 480, pp. 100–108, Nov. 2016, doi: 10.1016/j.jnucmat.2016.08.017.
[4] R. X. Li, P. K. Liaw, and Y. Zhang, "Synthesis of AlxCoCrFeNi high-entropy alloys by high-gravity combustion from oxides," *Materials Science and Engineering: A*, vol. 707, pp. 668–673, Nov. 2017, doi: 10.1016/j.msea.2017.09.101.
[5] X. H. Yan, J. S. Li, W. R. Zhang, and Y. Zhang, "A brief review of high-entropy films," *Materials Chemistry and Physics*, vol. 210, pp. 12–19, May. 2018, doi: 10.1016/j.matchemphys.2017.07.078.
[6] W. Li, et al., "Influence of Mo on the microstructure and corrosion behavior of laser cladding FeCoCrNi high-entropy alloy coatings," *Entropy*, vol. 24, no. 4, p. 539, Apr. 2022, doi: 10.3390/e24040539.

[7]   C.-Z. Yao, et al., "Electrochemical preparation and magnetic study of Bi–Fe–Co–Ni–Mn high entropy alloy," *Electrochimica Acta*, vol. 53, no. 28, pp. 8359–8365, Nov. 2008, doi: 10.1016/j.electacta.2008.06.036.
[8]   R. Chandrakar, A. Kumar, S. Chandraker, K. R. Rao, and M. Chopkar, "Microstructural and mechanical properties of AlCoCrCuFeNiSix (x = 0 and 0.9) high entropy alloys," *Vacuum*, vol. 184, p. 109943, Feb. 2021, doi: 10.1016/j.vacuum.2020.109943.
[9]   Y. Yao, et al., "Carbothermal shock synthesis of high-entropy-alloy nanoparticles," *Science (1979)*, vol. 359, no. 6383, pp. 1489–1494, Mar. 2018, doi: 10.1126/science.aan5412.

Index

https://doi.org/10.1515/9783110769470-011

www.ingramcontent.com/pod-product-compliance
Lightning Source LLC
Chambersburg PA
CBHW081545220326
41598CB00036B/6568